AGRICULTURAL BIOTECHNOLOGY

AGRICULTURAL BIOTECHNOLOGY

Issues and Choices

Information for Decision Makers

Edited by Bill R. Baumgardt and Marshall A. Martin

Purdue University Agricultural Experiment Station
West Lafayette, Indiana

This material is based upon work supported by the U.S. Department of Agriculture, Cooperative State Research Service (in cooperation with the Agricultural Research Service, Economic Research Service, and Extension Service) under Agreement No. 88-COOP-1-4181.

Any opinions, findings, conclusions, or recommendations expressed in this publication are those of the authors and do not necessarily reflect the views of the U.S. Department of Agriculture, the Committee on Biotechnology of the National Association of State Universities and Land Grant Colleges, or the editors.

Purdue University Agricultural Experiment Station,
West Lafayette, Indiana 47907

ISBN 0-931682-28-2

Printed in the United States of America

Contents

Preface ..... *vii*

Acknowledgments ..... *ix*

Chapter 1 **The Origins of Biotechnology and Its Potential for Agriculture** 3
Marshall A. Martin and Bill R. Baumgardt

Chapter 2 **Emerging Biotechnologies: Impact on Producers, Related Businesses, and Rural Communities** 23
W. Burt Sundquist and Joseph J. Molnar

Chapter 3 **Impact of Biotechnology on Food and Nonfood Uses of Agricultural Products** 41
Susan K. Harlander, James N. BeMiller, and Larry Steenson

Chapter 4 **Food Safety Issues in Biotechnology** 55
Michael P. Doyle and Elmer H. Marth

Chapter 5 **Environmental Risks and Benefits of Agricultural Biotechnology** 67
James J. Marois, James I. Grieshop, and L. J. (Bees) Butler

Chapter 6 **U.S. Biosafety Regulations: Too Much or Not Enough?** 81
D. R. MacKenzie and Anne K. Vidaver

Chapter 7 **Intellectual Property Rights** 95
V. L. Lechtenberg and A. A. Schmid

Chapter 8 **Agricultural Biotechnology: A Global Perspective** 107
Randolph Barker and Don Plucknett

Chapter 9 **Ethics and Values Associated with Agricultural Biotechnology** 121
Glenn L. Johnson and Paul B. Thompson

Chapter 10 **Public Perceptions of Agricultural Biotechnology** 139
William B. Lacy, Lawrence Busch, and Laura R. Lacy

Chapter 11 **Issues and Choices: How Will Agricultural Biotechnology Influence the Future?** 163
Neville P. Clarke and Charles E. Hess

Glossary 173

Authors 177

Reviewers 179

Preface

Biotechnology holds much promise as a new set of tools for increasing the efficiency and sustainability of production agriculture and assuring the safety, quality, variety, and quantity of food products. The techniques of recombinant DNA, tissue culture, gene transfer, embryo manipulation, and bioprocess engineering—among the other techniques that come under the heading of "biotechnology"—allow scientists to understand and manipulate life processes at the molecular level in ways that were unknown and unattainable just a few years ago.

But, as is the case with any technology, biotechnology does not come to agriculture without risk. In the early 1970s, when research involving recombinant DNA was first conducted, concerns were raised by the public and within the scientific community about the potential risks of this new technology. And today, sincere questions about agricultural biotechnology are still being raised by some people. These questions range from potential risks to human health, food safety, and the environment, to potential negative socioeconomic impacts on farming and rural communities, to concerns of an ethical nature.

In 1988, as part of its effort to take a proactive stance and address society's concerns about agricultural biotechnology, the Committee on Biotechnology of the Division of Agriculture of the National Association of State Universities and Land Grant Colleges commissioned its Socioeconomic Subcommittee to produce a basebook and a leaflet series that would discuss the potential socioeconomic impacts of agricultural biotechnology. Drs. Baumgardt and Martin, the chairpersons of the subcommittee, assumed the editorship of the combined project. Once funding from USDA was obtained, they identified appropriate chapter topics and potential authors.

In March 1989, an authors' workshop was held in St. Louis, Missouri. There, the authors shared their proposed chapter outlines and invited input from several discussants. These discussants, representatives from private industry, government agencies, academia, and public advocacy groups, critiqued the proposed outlines from their various perspectives.

With the benefit of this input, the authors wrote first drafts of their chapters and the complementary leaflets. At least two reviewers, selected from a cross section of the groups represented at the St. Louis workshop and with knowledge in the appropriate subjects, reviewed the first drafts of each chapter. Guided by the reviewers' comments and by editorial suggestions, the authors then prepared revised versions of their chapters.

The major purpose of the basebook, *Agricultural Biotechnology: Issues and Choices*, is to provide a reference source on the socioeconomic aspects of agricultural biotechnology for federal agency administrators, deans of agriculture, directors of agricultural research, Cooperative Extension Service directors, and other agricultural professionals. Such leaders and professionals should find in this basebook, concise, reliable background information from which to draw to upgrade their own understanding of the issues and with which to better prepare themselves to discuss these issues with the public and the media.

The leaflet series, under the same general title, is composed of briefer companion pieces to the chapters in the basebook. The leaflets can be used as handouts at public policy meetings, as student reading assignments, or as printed background material for the media and legislative aides.

The basebook covers three general topics related to agricultural biotechnology: food production and processing, risk and regulatory functions, and policy choices.

Chapter 1 offers some background on what biotechnology is and what it means for agriculture and for society in general.

Chapter 2 assesses some of the possible ramifications of agricultural biotechnology adoption for agriculture and for rural communities.

Chapter 3 examines the potential impact biotechnology can have on food and nonfood uses of agricultural products and the concomitant societal implications.

Chapter 4 discusses some of the food safety issues associated with agricultural biotechnology.

Chapter 5 addresses potential environmental benefits of and concerns about agricultural biotechnology.

Chapter 6 outlines the current status of agricultural biotechnology regulation, the regulatory functions of U.S. government agencies, and related regulatory issues.

Chapter 7 explains the history of intellectual property law as it has evolved in the United States, discusses how products of agricultural biotechnology fit into the law, and examines the global implications of intellectual property laws for the development of agricultural biotechnology.

Chapter 8 discusses some of the international ramifications of agricultural biotechnology.

Chapter 9 explores the diverse ethical concerns raised by agricultural biotechnology.

Chapter 10 identifies public concerns about agricultural biotechnology and summarizes the results of U.S. public opinion surveys on science and technology in general and agricultural biotechnology in particular.

Chapter 11 outlines the societal issues and choices concerning agricultural biotechnology and discusses how our decisions can shape the future.

Acknowledgments

Agricultural Biotechnology: Issues and Choices

A Basebook and Leaflet Series

On behalf of the Committee on Biotechnology of the National Association of State Universities and Land Grant Colleges, the editors express their appreciation to the many people who contributed to this basebook and leaflet project. It was truly a group effort of significant proportions.

First and foremost, we gratefully acknowledge funding for the project from the U.S. Department of Agriculture. The staff in the Cooperative State Research Service and the Office of Agricultural Biotechnology were extremely supportive of this effort, as were staff of the Office of Public Affairs and the Agricultural Research Service.

Second, we express our appreciation to the Socioeconomic Subcommittee of the NASULGC Committee on Biotechnology. This subcommittee defined the general scope desired in this project at a meeting in Washington, D.C. on March 31, 1988.

Our most sincere thanks go to all the authors whose professional expertise and dedication bring credibility to the basebook and leaflets. We appreciate their original ideas as they approached their respective writing assignments; their willingness to modify those ideas based on the vigorous, frank, and open discussion at an author's workshop held in March 1989; and their careful editing and rewriting during the review and publication process.

Special thanks go to Jean Rosscup Riepe, Research Associate in the Center for Agricultural Policy and Technology Assessment at Purdue University, for her constant attention to the many details necessary to successfully complete the project, and to Laura Hoelscher, Editor and Information Specialist in the Department of Agricultural Communication Service at Purdue University, for her tireless efforts as technical editor. These two individuals contributed above and beyond the call of duty. This basebook simply would not have been completed without their truly exceptional dedication to the project.

Many individuals in the Department of Agricultural Communications at Purdue University contributed generously to this effort. The editors would like to recognize the thoughtful creative and artistic contributions provided by Russell J. Merzdorf, who designed the covers and layout for the basebook and the leaflet series. We also acknowledge the expertise lent by Steven A. Cain toward the visual conceptualization of the ideas written by the authors. To the many others in the Department of Agricultural Communications and the Agricultural Experiment Station at Purdue University who contributed unselfishly with their special talents, we express our sincere thanks and ask that they please accept our apologies for not mentioning each of them by name.

Bill R. Baumgardt and
Marshall A. Martin

Photographs appearing in this volume are from a variety of sources. The editors gratefully acknowledge the kind permission to reproduce photographs granted by faculty and staff of Purdue University's School of Agriculture and by the USDA's Office of Public Affairs, Agricultural Research Service, and Cooperative State Research Service.

For information about ordering a copy of *Agricultural Biotechnology: Issues and Choices* or the companion leaflet series, contact:

Purdue University Agricultural Experiment Station
116 Agricultural Administration Building
Purdue University
West Lafayette, IN 47907

Text is printed on recyclable paper.

PRINTED WITH SOY INK

CHAPTER ONE

Agricultural biotechnology, the latest in the series of technological innovations that have transformed American agriculture, offers many opportunities and challenges. Many people are excited at the prospect of the benefits promised by agricultural biotechnology. Others are concerned that the potential risks outweigh those benefits.

Gregor Mendel's genetic research in the mid-nineteenth century laid the foundation for modern biotechnology. His work and that of later researchers have made it possible to apply the tools of biotechnology in agriculture.

Biotechnology is being applied in both plant and animal agriculture. In the plant area, the techniques of biotechnology are being used to facilitate plant breeding and to develop plant strains resistant to diseases, pests, and environmental stress. Basic research into the actual mechanics of DNA insertion in plants is also being conducted. Areas of animal research in which biotechnology is being applied include lactation and growth, reproduction, animal health, and genetics.

Many of the products of agricultural biotechnology will require government approval for commercialization, as is the case for other new products. Historically, products have been approved strictly on the basis of scientific criteria, with primary emphasis on efficacy, product safety, and environmental implications. However, the public is increasingly concerned about the socioeconomic ramifications of all new technologies. As a result, the broader societal issues related to agricultural biotechnology need to be assessed and understood by scientists and the general public alike if society is to make informed choices concerning the development and application of agricultural biotechnology.

1 The Origins of Biotechnology and Its Potential for Agriculture

Marshall A. Martin, Purdue University
Bill R. Baumgardt, Purdue University

Agricultural biotechnology is a new tool in the continuing evolution of agricultural technology in the United States. During the past century, agricultural mechanization swept the countryside and markedly increased labor productivity and efficiency. Following World War II, new hybrid seeds, fertilizers, and pesticides helped double, and even triple, crop yields. New vaccines, improved genetics and breeding, and better nutrition and management increased livestock productivity. Artificial insemination, refrigerated bulk tanks, and better nutrition management led to steady improvements in dairy cow productivity.

International cooperation in agricultural research and development blossomed after World War II, opening the doors for international exchanges of germplasm and information. This helped increase agricultural productivity in developed and developing countries alike.

As the twentieth century draws to a close, agriculture is entering a new technological era. The tools of modern biotechnology can be used to offer major new scientific insights into the functioning of plants and animals at the cellular and molecular levels. These discoveries promise new opportunities and challenges for food and fiber production in the United States and around the world, opportunities, previously undreamed of, to prevent plant and animal diseases, control insects without chemical pesticides, increase livestock productivity, enhance food quality, and reduce environmental degradation.

Yet, as with most new technologies, there is some apprehension and concern about the technical, social, and economic implications of the application of biotechnology to agriculture. This book outlines the nature and scope of agricultural biotechnology, addresses concomitant social and economic issues, and examines some of the alternative choices available to society regarding the impact and direction of agricultural biotechnology.

What Is Biotechnology?

Although the word "biotechnology" is relatively new, the subject has an ancient history. For thousands of years yeasts, molds, and bacteria have been used to make fermented foods like beer, wine, and bread and to preserve foods, such as by turning milk into cheese and yogurt. Biotechnology is around us every day, just as it was for our ancestors. But what was often mysterious to them can now be understood through the discoveries of biotechnology.

Biotechnology refers to methods of using plants, animals, and microbes, either wholly or in part, to produce useful substances or improve existing species. More specifically, biotechnology is the use of technologies based on living systems to develop commercial processes and products. The new biotechnologies include the techniques of recombinant DNA (rDNA), gene transfer, embryo manipulation and

Biotechnology Tools

- *recombinant DNA*
- *gene transfer*
- *embryo manipulation and transfer*
- *plant regeneration*
- *tissue culture*
- *monoclonal antibodies*
- *bioprocess engineering*

transfer, plant regeneration, tissue culture, monoclonal antibodies, and bioprocess engineering (National Research Council, p. 3). Biotechnology involves the integration of such advanced disciplines as biology (plant, animal, and microbial), biochemistry, molecular biology, genetics, chemical engineering, and computer science.

Agricultural biotechnology has its roots in food production. Today's agricultural industry will be a major beneficiary of modern biotechnology. Its use can result in the production of more suitable farm animals and crop plants, as well as improved foods and food-processing procedures. Dozens of examples of the application of biotechnology in agriculture are rapidly emerging.

Applications in Animal Agriculture

Successful embryo transplants are speeding up the genetic improvement of cattle at lower cost to the farmer. Embryo technology also offers an economically viable way of transferring superior animal genetics to developing or geographically isolated countries.

Several superior vaccines to prevent animal diseases have been developed through biotechnology. Also, disease diagnostic techniques using monoclonal antibodies are being developed to detect animal diseases, estrus and pregnancy, and aflatoxin in livestock feed.

In addition to controlling animal diseases, a major contribution of genetic engineering will be to increase the productivity of farm animals. For example, treating cows with supplemental bovine somatotropin (BST), a natural protein hormone manufactured through recombinant DNA technology, can increase milk production per lactation on less feed per unit of milk, thereby reducing milk production costs. Similarly, administering recombinantly produced porcine somatotropin (PST), a natural swine protein growth hormone, to finishing hogs can improve feed efficiency, increase lean muscling, and reduce fat deposition.

Applications in Crop Production

Biotechnology is opening new frontiers in crop and food production. Improvements in production efficiency and processing procedures should result in lower cost, higher quality foods in the years ahead.

Currently, there are many applications of biotechnology in plant research, including tissue culture techniques and the genetic engineering of single-gene traits. Tissue culture, which is the maintenance and growth of plant cells in the laboratory, is being used to produce identical copies (clones) of plants with superior genetics. Genetic engineering has made the identification and transfer of single genes into certain plants a relatively common practice. Whole plants have been regenerated from cultured cells that have been genetically altered.

Biotechnology offers alternatives for controlling plant pests, such as insects, weeds, and diseases, based on biological rather than chemical methods. Scientists employing genetic engineering techniques have inserted into plants the gene for Bt-toxin (from the bacteria *Bacillus thuringiensis*), making the plants resistant to some insects. Scientists have identified the plant genes that cause resistance to several herbicides. Incorporating these genes into field crops may reduce herbicide costs and allow farmers to use more environmentally benign chemicals.

Another area of potential crop improvement through biotechnology is the regulation of plant growth and development. Genes responsible for hormones that control plant growth and development processes have been identified. These discoveries offer, for instance, longer shelf life for fresh vegetables and flowers.

Other Applications of Biotechnology

Biotechnology can contribute to more efficient food processing and higher quality foods. Current research includes modifying field crops to improve their nutritional value or processing characteristics. Through biotechnology, scientists have been able to make more efficient use of enzymes, the proteins that catalyze chemical reactions. Enzymes are widely used in bioprocessing. Researchers also are applying biotechnology to transform food-processing wastes into valuable products.

Biotechnology also is being applied to expanding nonfood uses of farm commodities and improving tree genetics. Altering the structural characteristics of starches and oils in crops may result in new or wider industrial applications. The difficulty in breeding trees, caused by their extended life cycle, can be circumvented through biotechnology.

Expected Benefits and Risks of Biotechnology

Basic and applied research in the various disciplines, drawing from biotechnology, offer opportunities for increasing agricultural productivity and profits, lowering food costs, and maintaining the competitiveness of U.S. agriculture in world markets. However, there are concerns about potentially adverse socioeconomic effects of agricultural biotechnology. Biotechnology is a controversial topic because of its potential to reshape the world in which we live. Various groups in society approach biotechnology from different perspectives based upon how they anticipate biotechnology may alter their world.

The application of agricultural biotechnology will affect several areas of life, as well as many groups in society and their individual members. As with any new idea or technology, changes will be necessary in attitudes, management approaches, and industrial structure and organization. These changes, and the challenges they will present, will be exciting and advantageous for some, while threatening and possibly disastrous for others. Individuals view differently the potential benefits and risks that agricultural biotechnology may bring. A discussion of the areas of concern follows.

Science

Biotechnology is generating a great deal of excitement in the scientific community. Biotechnology tools are facilitating basic research efforts to understand the intricate, complex, and often mysterious workings of living organisms. Specific biotechnology applications often necessitate further basic research. The basic research being conducted in conjunction with biotechnology is accelerating the accumulation of knowledge in traditional disciplines such as biology, genetics, plant physiology, and biochemistry.

Increased understanding of genetics and other basic processes is fueling the application of biotechnology in industries such as agriculture which depend on living systems. Agricultural scientists in many disciplines are applying gene identification, genetic engineering, embryo technology, and tissue culture techniques to develop new products

Biotechnology is generating a great deal of excitement in the scientific community.

and processes. Biotechnology tools often complement and extend—rather than replace—traditional methods used to enhance agricultural productivity and to develop new production systems.

Biotechnology offers scientists the necessary tools to study the interactions among crop plants, weeds, soil organisms, bacteria, and the whole host of living organisms involved in production agriculture. Better understanding of these biological interactions is expected to result in the development of farming systems that are more sustainable in the long term and less dependent on costly, and potentially hazardous, inputs.

Scientists are aware of possible risks associated with biotechnology. Product and safety testing are important. Hence, strict research guidelines have been established, especially for field testing of genetically altered plants and microorganisms and for product testing of recombinantly produced food ingredients and animal health or growth products.

Adequate funding for research concerns many scientists. There is the potential for a disproportionate amount of agricultural research funding being directed towards biotechnology to the detriment of traditional disciplines. Yet, underfunding of agricultural biotechnology research is a more likely scenario in an era of federal budget deficits and given that society historically has underinvested in agricultural research (Griliches).

Agricultural Production

Farmers and ranchers are anticipating increases in the efficiency of crop and livestock production from the application of biotechnology. This will result in lower production costs and increased economic profits per unit of output, especially for the innovative early adopters.

New inputs and products and alterations of traditional inputs and products are expected from biotechnology. Some genetically engineered inputs, such as insect-resistant plants that allow farmers to apply fewer or less dangerous insecticides, will substitute for traditional inputs in the production process. Farmers will continue to purchase most of their inputs from agribusiness firms, but the mix of purchases will change. Farmers will require more support services and technical information to use correctly the biotechnology products, e.g., crops genetically engineered for insect resistance or BST to promote more efficient milk production. New products may arise through biotechnology as scientists introduce new crops for commercial production, adapt traditional crops to wider geographical locations, or alter the characteristics of traditional crops or livestock.

Biotechnology will lead to increased farm production efficiency, but this may result in a need for better management skills.

Biotechnology should not be perceived as a panacea that will solve all farm production and management problems. In fact, management skills will become increasingly important in many instances as producers strive to maintain and interpret more detailed financial and production records, often computer based. New or altered traditional products may mean developing new markets or marketing skills. Farmers who are unable or unwilling to upgrade their management skills may not remain cost competitive, and some will likely leave farming, as have many in past decades as a consequence of technological change.

Food Prices and Safety

Consumers ultimately will be the primary beneficiaries of agricultural biotechnology. As food production and processing costs per unit decline, depending on market structure and marketing margins, a portion of these cost reductions will be passed on to consumers in the form of lower retail prices. In addition, biotechnological developments will enhance the availability, diversity, quality, and convenience of retail food products.

Biotechnology raises concerns for some, particularly concerning food safety. The United States, through the

Food and Drug Administration (FDA) and other government regulatory agencies, has in place one of the best systems in the world to monitor and regulate food safety in the production, processing, and distribution of food. Still, occasional errors or accidents and inadequate or misleading information on new products, food additives, or pesticide residues have resulted in heightened consumer sensitivity to food safety issues. Biotechnology will bring about changes in food production and processing methods and in the chemical composition and nutritive value of foods. Any changes will require careful monitoring and consumer education to avoid unsafe food products or unwarranted consumer concerns.

One of the biggest challenges to the development and application of agricultural biotechnology will be establishing a dialogue with the consuming public to minimize unwarranted fears and provide adequate information so that consumers can make informed choices. The success or failure of agricultural biotechnology will hinge on consumer acceptance of products derived from it.

Biotechnology promises to aid efforts to preserve and enhance our natural resources and environment. The potential risks that do exist are being dealt with through the establishment of guidelines and the careful monitoring of research and field testing.

Agribusiness

Agribusiness firms are investing millions of dollars in basic and applied research in biotechnology to discover, develop, and market new products. Some veterinary products already have been approved by the FDA, while other biotechnology products, such as BST and PST, await FDA approval. With increased regulatory pressures and public concerns about environmental quality and food safety, biotechnology may enable agribusiness firms to offer new products that will increase food production efficiency while minimizing the environmental degradation associated with the use of some agricultural chemicals.

While agribusiness firms expect to gain new markets and revenue sources, they are facing several obstacles to new product development. Regulatory restrictions continue to expand, requiring further testing, review, and approval of new products. Excessive restrictions result in unnecessary costs and delays in releasing a new product. Furthermore, despite careful testing and review, public controversy can further delay the approval of a product for commercial use. A current example is the debate surrounding the FDA approval of BST for commercial use by dairy farmers (Fallert, et al.).

The Environment

During the past two or three decades, many people have become increasingly aware of and concerned about society's use and abuse of natural resources and the environment. Population growth, industrial development, and modern farming systems generate needs for industrial and agricultural production practices that sometimes require chemicals or procedures that may cause water or air pollution or soil erosion. The current emphasis on sustainable, alternative agriculture recommends the reduction in pesticide use, the adoption of conservation tillage methods, and the incorporation of legumes into crop rotations. Many farmers already have adopted these practices and others, including integrated pest management (IPM) techniques. Yet, more must be done to ensure an adequate, safe food supply at reasonable prices without the environmental degradation that has been associated with some modern agricultural practices.

Biotechnology promises to aid efforts to preserve and enhance our natural resources and environment. Insect- and herbicide-resistant plants and grasses with nitrogen-fixing capabilities (as possessed by legumes) may help farmers use fewer pesticides and less commercial nitrogen fertilizer, reducing potential environmental degradation from current farming systems. Biotechnology can help scientists better understand the interactions of living organisms in farming systems, enabling them to develop alternative farming systems that work more effectively with the environment.

While recognizing the potential benefits of biotechnology, some environmental groups are voicing concerns over its use. They fear that the use of herbicide-resistant crops will lead to greater—not less—herbicide use. Others question whether it is safe to release genetically altered species into the environment. To avoid potential risks of genetic engineering, carefully designed research guidelines and procedures have been developed and are being followed for field testing of genetically altered microorganisms, livestock, and plants (Purchase and MacKenzie).

Ethics

Some people express ethical concerns about biotechnology. They often raise difficult philosophical and ethical questions. Is the transferring of genes across species morally defensible? Should we tamper with God's creation? Do we have the right to replace "bad" genes with "good" genes? Who decides?

Others worry about the concentration of the agricultural input industry into a few firms responsible for most of farmers' inputs, e.g., seeds, fertilizers, animal health products, and pesticides. They fear that industry concentration will result in market power and price discrimination that may limit farmers' access to desired inputs and/or increase their costs above competitive market levels.

This ethical questioning is not unique to biotechnology. Traditional plant and animal breeding, adoption of new agricultural practices (e.g., monocropping, artificial insemination, and livestock confinement), introduction of new foods (e.g., oleo, pasteurized milk), and structural change in agriculture have been occurring for several decades. People have questioned the wisdom and morality of all of these practices or products at one time or another, and some continue to do so. Biotechnology may accelerate the rates of technological adoption and structural change in farming and other agricultural businesses and thus raise new questions.

Government Spending

The U.S. federal budget deficit remains high by historical standards, and the federal debt continues to grow. Thus, taxpayers are concerned about any policies that may increase government spending. Gains in agricultural productivity in the past have often resulted in commodity surpluses and declining market prices. Consequently, farm programs have been implemented to stabilize commodity prices, regulate stocks, and support farm income. These policies cost taxpayers money. Part of the recent controversy with BST is a concern that excessive milk supplies will be generated by increasing milk productivity per cow, resulting in depressed farm prices, increased government purchases of dairy products, and increased U.S. Treasury outlays.

Historically, the cooperative federal-state land-grant system has invested tax dollars in agricultural research and extension activities. These investments have increased agricultural efficiency, generated new jobs, and increased profits for many, such that the social returns to agricultural research have far outweighed the costs. The challenge is to compare the expected taxpayer costs of biotechnology research support with potential benefits, which include increased profits to farmers and agribusiness firms; lower priced, higher quality food to consumers; a reduction in environmental degradation; and an improved competitive posture in international markets.

International Interests

International competitiveness is a key national policy goal today. U.S. agriculture during the past quarter century has become an important source of foreign exchange earnings, helping reduce the U.S. trade deficit. In recent years the United States annually has exported about 150 million metric tons of products valued at approximately $40 billion. These exports represent production from about two out of every five crop acres. Farmers in other countries also are adopting new output-increasing, cost-reducing technologies. Thus, to remain competitive in world markets, the United States must increase agricultural productivity and marketing efficiency. Biotechnology can help achieve these goals.

The United States, however, cannot easily control biotechnology developments in other countries nor the international flow of scientific information. The international agricultural research community, through conferences, professional journals, computer networks, and other methods of communication, is constantly exchanging ideas, techniques, and germplasm.

Some people are concerned that too much information will flow out of the United States, threatening our international market competitiveness. Others perceive that impoverished nations may be excluded from the transfer of biotechnology information and, as a result, from the benefits of biotechnology. This raises concerns in the minds of some about the control of and access to biotechnology discoveries and about who will benefit from them. Intellectual property rights is one issue that can be negotiated in international forums—an example is the General Agreement on Tariffs and Trade (GATT).

Agricultural biotechnology clearly offers exciting opportunities for society. Yet, many issues and concerns must be faced and decisions made by scientists, legislators, regulatory agencies, producers, business executives, and consumers. The challenge of effectively using biotechnology to achieve society's goals will require constant efforts to conduct objective research, share information, and make decisions collectively.

Genetics and the New Biotechnologies

Biotechnology is the new ability, courtesy of genetic engineering, to modify the blueprints of heredity of every living creature, plant, or organism. For many, genetic engineering is synonymous with biotechnology. Thus, a digression into the evolution of the tools of biotechnology and genetic engineering may help to clarify and demystify the role of these tools in the advancement of agriculture.

The Birth of Genetics

The foundation of modern genetics began in 1865 when Gregor Mendel presented his laws of heredity to the Natural Science Society in Brunn, Austria. Mendel crossed pea plants of differing morphologies and not only noted the appearance of specific traits in the first generation of progeny, but also went on to study crosses among these progeny, as well as crosses of the progeny and the original parent pea plants. He concluded that many traits of the pea were under the control of two distinct "factors," one from the male and the other from the female parent. Further observations included the concepts of phenotype versus genotype and dominance versus recessiveness.

Mendel died in 1884 after 41 years of study, with no scientific acclaim for his landmark experiments on the hereditary "factors" of pea plants. He said not long before his death, "My time will come." He was correct in that the science of genetics was reborn when Mendel's work was rediscovered in 1900 by three scientists, Hugo DeVries, Erich Von Tschermak, and Carl Correns, each independently checking the scientific literature for precedents to their own "original" work.

The term "genetics" was not actually coined until 1906, and it was 1909 before the term "gene" was used to replace Mendel's term "factors." At about the same time that Mendel made his observations, it became known that hereditary factors are transmitted through the egg and sperm. Soon afterward, as it became possible to

Mendel died in 1884 after 41 years of study, with no scientific acclaim for his landmark experiments on the hereditary factors of pea plants.

visualize the chromosomes within the nucleus, it was conjectured that the function of the nucleus is to carry the hereditary information. Cells of individuals of a given species were found to contain a constant number of chromosomes. Further investigation showed that chromosomes behaved exactly as they should if they were indeed bearers of the genes that Mendel described. As many more genetic experiments were conducted, it became apparent that a large number of different genes are located on each chromosome.

The Discovery of DNA

Modern molecular biology was not possible until deoxyribonucleic acid (DNA) was discovered. The history of DNA began with Friedrich Miescher, a Swiss biologist who, in 1868, isolated nucleic acid from pus cells obtained from discarded bandages. He later identified a similar substance in the heads of salmon sperm cells. Although he separated the nucleic acid fraction and studied its properties, the covalent structure of DNA did not become known until the late 1940s.

It was not until 1943 that direct evidence was provided that DNA is the bearer of genetic information. Oswald Avery, Colin MacLeod, and Maclyn McCarty, working at the Rockefeller Institute, discovered that DNA taken from a virulent strain of bacterium permanently transformed a non-virulent form of the organism into a virulent form. They concluded it was the DNA from the virulent strain which carried the genetic message for virulence and that it became permanently incorporated into the DNA of the recipient non-virulent cells. Although the scientific community was slow to adopt the idea, it is now accepted that DNA carries the genetic information from one generation to the next in all living cells; thus, DNA has been described as the "blueprint for life."

The next key step was to determine the mechanism by which DNA passes genetic information on to the next generation. In 1953 scientists James Watson, an American geneticist, and Francis Crick, an English physicist, working at the University of Cambridge, proposed a double helical structure for DNA. This was the culmination of a series of investigations by a number of scientists. Watson and Crick were able to assemble the information like putting together the pieces of a puzzle to produce their model of the structure of DNA. For their landmark work in discovering the double helical structure of DNA, Watson and Crick shared the 1962 Nobel Prize for Physiology and Medicine with Maurice Wilkens. Another contributor to the work, Rosalind Franklin, had died the year before and thus was not eligible to be named as a recipient.

Although the scientific community was slow to adopt the idea, it is now accepted that DNA carries the genetic information from one generation to the next in all living cells; thus, DNA has been described as the blueprint for life.

The model developed by Watson and Crick accounted for the periodicities along the axis of the DNA fibers, as well as the specific adenine (A)/thymine (T) and guanine (G)/cytosine (C) bases discovered earlier by others. Their solution involved building a three-dimensional model of the structure of DNA consisting of two helical chains of DNA coiled around the same axis to form a right-handed double helix; the two strands of DNA run in opposite directions and are complementary, being matched by the hydrogen bonds of the A-T and G-C pairs. When DNA replicates, an exact duplicate of the parental genetic information is made, due to the pairing of the bases.

Gene Action and Movement

Discovery of "inborn errors of metabolism" as early as 1909 by English physician Archibald Garrod led to the speculation that since key enzyme(s) were missing, then for every enzyme, and perhaps for each protein that a cell possesses, there must exist a corresponding gene. Studies in the 1940s by Stanford University geneticist George

Beadle and the biochemist Edward Tatum on the mold *Neurospora* verified the one-gene-one-enzyme concept. Identification of the double helix concept made it possible to develop more precise information about the one-gene-one-protein relationship. Classic work on the genetics of *E. coli* begun in the 1940s by Seymour Benzer at Purdue University and by Joshua Lederberg and Edward Tatum at Yale University (who first sexually crossed mutant bacteria) greatly extended the knowledge of gene action.

Various organs or groups of cells within an animal or plant are quite different in their structure and composition in spite of the fact that the DNA of every cell in the organism is identical. In different cell types, different regions of the DNA are "active." An understanding of this difference in DNA activity in different parts of the plant or animal involves what is often known as the central dogma of molecular biology: DNA makes RNA makes protein. As considered by the molecular biologist, a gene is that portion of DNA that encodes for a single protein. The strict idea of "one gene makes one protein" has required some modification with the discovery that some proteins are composed of several different polypeptide chains, but the "one gene makes one enzyme" rule does hold.

In the early 1950s, while working on the corn plant at the Cold Spring Harbor Laboratory, Barbara McClintock began to call attention to the novel behavior of genetic fragments she called "controlling elements." These genetic elements were first noticed by McClintock, sometimes referred to as a modern-day Mendel, because they inhibited the expression of other corn genes with which they came into close contact. They did not have fixed locations. Instead, they appeared to move around the corn genome. Thus, they also have been called "jumping genes." The elements could be inserted and removed. After removal, the function of a previously dormant gene often reappeared. When genes "jump," they can cause changes in the actual genes associated with them (mutations), as well as in the arrangement of genes along the chromosomes. Either of these can change which genes are expressed, and thus which characteristics become manifested. It is now suspected that transposable elements are a major feature of all DNA. Although it took over thirty years for her work to be recognized, in 1983 at the age of 82, Barbara McClintock became the first American woman to win the Nobel Prize individually.

Steps in and Uses of Genetic Engineering

Gene transfer, or recombinant DNA (rDNA) technology, is a key element in modern biotechnology. Essential steps in rDNA, or "genetic engineering," include: identification of a particular gene that encodes for a desired trait in an organism; isolation of the gene; study of its function and regulation; modification of the gene by altering the nucleotide sequence; and introduction of the hybrid DNA into a selected cell for reproduction and synthesis.

The first step in an effort to genetically engineer an organism is to locate the relevant gene(s) among the thousands that make up the genome. Isolation is made easier with "restriction enzymes" that can cut complex macromolecules of DNA into manageable pieces. A restriction enzyme recognizes a unique sequence in the DNA and cuts the strands. By using a series of restriction enzymes, DNA can be cut in lengths of one or several genes. The smaller segments can be sorted and then cloned. Thus, a gene library can be constructed for a particular genotype.

Hamilton O. Smith and Daniel Nathans received the Nobel Prize in Medicine in 1978 "for discovery of restriction enzymes and their application to the problems of molecular genetics." Herbert Boyer, Stanley Cohen, and colleagues developed the first practical methods for systematically cloning specific DNA fragments. Among other recent Nobel Prize winners for work in molecular biology are Paul Berg in 1980, for constructing the first recombinant DNA molecules, and Walter Gilbert and Frederick Sanger in 1980, for independently developing DNA sequencing procedures.

Recombinant DNA provides additional tools, beyond those of conventional genetics, for use in genetic improvement. Advantages include easier manipulation of the genome, circumvention of incompatibility between species, and more rapid genetic change. Additional basic research is needed to target foreign genes to specific sites in the host genome and to understand the control or expression of the foreign gene in plants and animals.

Concerns about the safety of certain recombinant DNA experiments were raised in the early 1970s. Scientists gathered at Asilomar, California, for the first international conference on the potential dangers of recombinant DNA.

In an act of responsible self-regulation, they recommended that guidelines be placed on their work. This resulted in the "NIH-RAC" (National Institutes of Health-Recombinant Advisory Committee). Responsibility and safety prevailed, so that in 1979 the NIH guidelines were relaxed. This scientist-driven guideline approach remains effective as a trustworthy element for the new technologies.

Biotechnology Applications in Plant Science

Biotechnology offers several techniques to improve plant genetics and breeding. Advances in biotechnology applied to plants include DNA insertion; improvements in disease, insect, and herbicide resistance; environmental stress tolerance; nitrogen fixation; and photosynthesis.

The Mechanics of DNA Insertion

Mechanisms of transferring genes and controlling their expression to achieve the desired result represent the most serious challenges for scientists working in plant and farm animal biotechnology. Transferring genes from one organism to another (transformation) requires the use of a vector, or carrier, to introduce the new DNA into a single cell, which can then multiply, carrying the new information to each new cell.

Vectors include viruses, plasmids, and transposable elements. Alternates for vector systems include laboratory means, such as chemical treatments; electroporation; and physical treatments, such as microinjection. Commonly used vectors in plants include a piece of the Ti plasmid (T-DNA) from the soil bacterium *Agrobacterium tumefaciens*, which has been genetically modified by eliminating its tumor-forming properties, and the Ri plasmid of *A. rhizogenes*. The Ti and Ri plasmids can transfer DNA only to the nucleus of dicotyledonous species, but not to the mitochondria and chloroplasts, which govern energy production and photosynthesis. Introduced traits have been found to be stable over at least five generations on genetically engineered tomato and oilseed rape plants.

Success with the Ti system has been possible for some time with plants such as petunia, tomato, potato, and tobacco, and now has been expanded to include most broadleaf crops. However, few monocotyledonous plants appear to be natural hosts for *Agrobacterium*. Cereal grains such as corn, wheat, and rice have not been successfully transformed by *Agrobacterium*. Therefore, extensive efforts have been directed toward the development of systems for the delivery of free DNA into these species.

Transformation has been achieved in plant protoplasts through facilitation of DNA uptake by calcium phosphate precipitation, polyethylene glycol treatment, electroporation, or combinations of these treatments. The usefulness of these methods is limited by difficulties involved in regenerating plants from protoplasts. Success has been achieved with rice, but the response with corn is as yet unreliable, although reports appeared in 1989 on the regeneration of fertile corn plants from protoplasts. In these cases, success is limited to certain lines, and commercialization will require a lengthy period of backcrossing. However, it should be only a short time before the full potential of these results is realized.

Gene gun technology is one method of transferring foreign DNA into intact plant cells or tissue, which can then be easily regenerated into plants.

Some scientists have been working on alternate ways to introduce DNA into intact cells or tissues. Regeneration of cereals from immature embryos or from explants, as opposed to protoplasts, is relatively simple. One of the most interesting approaches has been the introduction of a "particle gun," or high-velocity microprojectile technology. In this system, DNA is placed on the surface of small metal particles, which are then injected (shot) at high speeds through the cell wall and into the cytoplasm. These fast-moving particles are capable of moving through several layers of cells and permit the transformation of cells within tissue explants. Since this approach eliminates the need to regenerate plants all the way from the protoplast stage, the particle gun method has the potential to allow direct transformation of commercial lines of cereal plants.

One other major difficulty in genetically engineering plants is that many of the more important economic traits in plants are controlled by multiple genes. Current genetic engineering techniques are most effective when dealing with one gene at a time. Continued research on the basic genetic engineering techniques and on gene mapping should eventually lead to multiple-gene transformations.

Interaction of Biotechnology and Plant Breeding

Methodological challenges such as those discussed above provide one reason why the new biotechnologies will not replace classical plant breeding. In most cases, significant progress will involve teamwork between scientists using genetic engineering and those using more traditional plant breeding and genetics techniques.

Plant breeders have traditionally improved plant varieties by selecting on the basis of phenotype. One biotechnology tool that appears especially promising is the development of high-resolution restriction fragment length polymorphism (RFLP) genetic maps. Mapping with RFLPs involves the application of molecular biological techniques to the basic concepts of transmission genetics and was first described in 1980 by researchers in human genetics. These maps provide a more direct method for selecting desirable genes via their linkage to easily detectable RFLP markers. This technique is being used to map plant genomes. A version of this technology is "fingerprinting," which involves taking advantage of small segments of DNA that are dispersed throughout the genome and which are specific to the individual within a species. Major efforts are underway around the world in both the public and private sectors to develop new RFLP probes and genetic maps for many important crops. This new technology can be used in the evaluation and selection of crop materials for breeding purposes.

Biotechnologies, including tissue culture and micropropagation, also are being rapidly developed for use in forest tree improvement. Tissue culture should contribute significantly to the improvement of forests through exploitation of existing genotypes and production of new, commercially valuable genotypes.

Genetic Changes in Economically Important Plants

Biotechnology offers opportunities to improve economically important plants in a variety of ways:

- increased insect and disease control;
- new approaches to weed control;
- improved resistance to environmental stresses such as drought, cold, salt, or toxic elements such as aluminum;
- more efficient use of nutrients such as nitrogen;
- improved product quality through changes in characteristics such as fatty acids, amino acids, flavor, texture, and the ability to retain quality while in storage; and
- improvements in plant metabolic characteristics such as nitrogen utilization, photosynthesis, flowering and seed set, and the partitioning of nutrients between grain and stalk.

Disease and Insect Control

Conventional plant breeding, especially in agricultural experiment stations and in USDA-ARS, has emphasized breeding disease and insect resistance into economically important plants for many years. Biotechnology provides an additional set of tools to aid in these efforts. Genetically resistant plants minimize the need for chemical treatments and thus have both economic and environmental advantages.

Tobacco mosaic virus (TMV) has been a widely studied system. Nearly 60 years ago scientists learned that a mild strain of TMV could protect tobacco plants against the adverse effects, when subsequently inoculated with severe strains of the virus. There are potential problems with this conventional cross-protection approach because the mild, protecting virus might spread to other crops or mutate to a more virulent form.

Genetically resistant plants minimize the need for chemical treatment and thus have both economic and environmental advantages.

Recently, biotechnologists induced significant resistance to TMV infection by expressing only the coat protein DNA of TMV in transgenic plants, termed "coat protein-mediated protection." Similar results have been reported in transgenic tomato, tobacco, and potato plants against a broad spectrum of plant viruses. Because these "transgenic" plants have only a portion of the genetic information that is needed for TMV replication, the problems of conventional cross-protection are avoided. It appears that virus resistance introduced by recombinant DNA technology can be transmitted through seed as a simple Mendelian trait and can thus be transmitted by conventional breeding techniques.

Disease diagnosis and screening plants for disease resistance represent other useful applications of agricultural biotechnology. DNA probes, isolated single DNA strands used to detect the presence of the complementary strands, can be used to detect plant viruses and viroids. Detection permits rapid diagnosis and screening to halt the spread of disease.

Plant biotechnology research in insect resistance also is progressing. Transgenic plants with insect resistance have been obtained through the use of the insect-control protein genes of *Bacillus thuringiensis* (Bt), which has been marketed as a biological insecticide for more than 20 years. The gene for the protein toxin has been cloned. Now, biotechnology is being used to transfer the gene from the bacteria into plants to make them self-protecting. This has been done successfully with tomato, cotton, and tobacco plants, which resisted attacks from caterpillars in field trials. Early market opportunities for caterpillar resistance are leafy vegetable crops, cotton, and corn. Crop targets for beetle resistance are potato and cotton.

Other types of insecticidal molecules are being researched to extend biotechnology for controlling additional insect pests in target crops. For example, a strain of the corn root-colonizing bacteria called *Pseudomonas fluorescens* has been genetically changed so it produces an endotoxin that is a potent insecticide for certain pests, including black cutworm. The recombinant bacterium can be freeze-dried and coated directly on seeds before planting, or it can be sprayed on fields. Although the product currently has a narrow range of target insects (it shows toxicity to root cutworm but not to corn rootworm, a more important pest), it is being used as a prototype for further development.

From quite different studies, insect genes that fight bacterial infections by causing the production of antibiotics have been isolated and cloned. These genes are being transferred into potato plants since several bacterial diseases of potatoes are susceptible to the insect antibiotics. Another example of success in insect resistance is the report of a genetically engineered cotton plant that is highly resistant to bollworm, tobacco budworm, and beet armyworm. Research in engineering resistance to fungal pathogens and to bacteria remains in the early stages.

Weed Control

The use of biotechnology to improve weed control is being pursued on several fronts. One method is to introduce herbicide resistance into crops so that environmentally safe herbicides can be used. Another is to introduce herbicidal properties into crops themselves. In either case, the use of external chemical herbicides can be minimized.

Interest in developing herbicide-resistant plants was stimulated by the desire to make forest trees resistant to herbicides so that better weed control could be practiced and reforestation accelerated. Another interest is related to some newer herbicides which combine high effectiveness with nontoxicity to animals and rapid degradation in the soil. However, these herbicides often lack selectivity, which has limited their use to preemergence applications. Classical breeding methods have met limited success in developing herbicide-resistant cultivars.

Two general approaches have been taken in using biotechnology to obtain herbicide resistance: (a) altering the level and sensitivity of the target enzyme for the herbicide and (b) incorporating a gene that will detoxify the herbicide. Emerging technologies which show promise include *in vitro* mutagenesis and/or selection, para-

Biotechnology may contribute to overcoming environmental stress problems such as drought.

sexual hybridizations, embryo rescue, and, ultimately, genetic transformation. Here again is a case where conventional breeding techniques very likely will have to be linked to biotechnology to produce successful, complete systems of crop improvement. In some instances, single-gene resistance has been identified, making the problem easier to solve. For example, the gene responsible for resistance to glyphosate (Roundup®) has been inserted into tomatoes, tobacco, soybeans, cotton, and poplar trees.

Cell-culture techniques (i.e., *not* rDNA) have been used successfully to develop a corn plant with natural resistance to the imidazolidinone family of herbicides (Scepter,® Pursuit®). Several laboratories are pursuing resistance to sulfonylurea herbicides (Glean,® Oust,® and Ally®), as well as to bromoxynil (Buctril,® Brominal®) and atrazine. Research is now underway to incorporate this trait into hybrid corn lines, and commercialization has been predicted for the near future.

Environmental Stress Tolerance

Problems associated with environmental stresses such as drought, heat, cold, ozone, salt, and toxic minerals in the soil may account for as much as 80 percent of the difference between potential and realized crop yields. Biotechnology may contribute to overcoming some of the regional production disadvantages due to environmental characteristics (e.g., climate, soil). Thus, the location of agricultural production could be based more on transportation costs and marketing considerations than on climate. Both biotechnology and conventional plant breeding techniques are being used to introduce stress resistance. For example, plants with natural drought resistance, identified through tolerance screening procedures, are being crossed with commercial lines of several crops. Traits being pursued include improved root structure, metabolic adjustments, and temporary dormancy.

Frost protection has been approached by genetically altering microorganisms that interact with plants. A strain of the bacteria *Pseudomonas syringae* is normally found on plant surfaces, especially leaves and fruit. Most strains of this bacteria produce a protein around which ice crystals form at temperatures slightly above freezing. The crystals rupture plant cells, causing frost damage. By deleting the gene responsible for this protein and spraying the altered bacteria on plants, frost forms on the plants only at lower temperatures. Field trials with a preparation of this "ice-minus" organism (Frostban®) have been conducted successfully on strawberries and potatoes.

Nitrogen Fixation

Nitrogen (N) is one of the key nutrient needs for plant production. Whereas animals get their N by eating plants, plants themselves must obtain N directly from inorganic sources. As with microbes, plants demand a "fixed" source of nitrogen, such as nitrate or ammonium ions, rather than the "unfixed" form present in the air. Many leguminous plants, such as soybeans, peas, beans, peanuts, and alfalfa, use colonies of bacteria to extract, or "fix," nitrogen from the air.

On the other hand, the majority of the world's major crop plants, such as corn, wheat, and rice, do not have built-in symbiotic nitrogen-fixing bacteria. Nitrogen fixation is a very complex phenomenon, and, while transference of this trait to non-legumes is being researched, it will be a long-term, difficult problem to solve. Some progress has been made, largely because of the body of knowledge available from many years of basic research on the genetics and biochemistry of N-fixation.

Efforts to transfer the nitrogen-fixation trait are underway with both symbiotic N-fixing bacteria (*Rhizobium*) and free-living N-fixing bacteria (*Klebsiella*). Nitrogen fixation is carried out by the enzyme nitrogenas; and a set of about a dozen genes, the nif genes, are needed for assembly of the biochemical machinery of fixation. Researchers have identified bacterial genes, the nod genes, involved in nodulation. When bacteria invade leguminous plants, the nod genes are activated, nodules form where the bacteria reside, and nitrogen fixation begins. Research is in progress using several different strategies. One is to genetically engineer the *Rhizobium* to fix N more efficiently in their natural host legumes. Another approach is to find ways of making *Rhizobium* effective in cereal plants. Still another approach is to transfer the N-fixing ability to other bacteria that already live in association with a plant, but which do not now fix N. Still, some scientists hope to engineer plants to fix N themselves.

Photosynthesis

Photosynthesis, the complex process by which green plants convert sunlight to energy, possibly could be made more efficient, thereby allowing plants to grow faster. Much research has been done on the genes which encode for the enzyme ribulose-1,5-bisphosphate carboxylase-oxygenase (Rubisco), the key catalyst in photosynthesis. Rubisco consists of eight large protein subunits encoded by genes in chloroplasts and eight small subunits encoded by genes in the nucleus. At the present time, no vector system exists to alter genes in the chloroplasts, but it appears that the high-velocity particle system will work. Success in this area likely is some years away (similar to N-fixation), but researchers hope to produce plants that will grow faster by increasing the efficiency with which Rubisco fixes atmospheric carbon dioxide.

Biotechnology Applications in Animal Agriculture

Biotechnology offers new opportunities for improving farm animal production efficiency and qualitative features of animal products. Advances in biotechnology are resulting in improvements in the areas of lactation and growth, reproduction, health, and genetics.

Lactation and Growth

The somatotropins (or growth hormones) are likely to be among the first products of biotechnology to have significant impacts on the productivity of animal agriculture. Somatotropin, a protein (*not* a steroid) hormone secreted naturally by the pituitary gland in mammals, directs lactation and growth.

Many of the effects of somatotropin have been known for years, but research with large animals was limited when the only source of the somatotropin was extraction from pituitary glands of humans or farm animals. Each species produces its own type of somatotropin, and thus it is not generic across species. Furthermore, the somatotropin produced by one species is only active in that species or a "lower" order species, not a "higher" order one. For example, bovine somatotropin (BST) is active in cattle and mice, but not in humans.

Many of the functions of this hormone were discovered in the late 1930s when scientists learned that it naturally causes increased growth rate, enhanced mammary gland growth and development, and increased milk production. During World War II, BST use was considered as a method to increase milk production in England, but not enough pituitary gland material could be obtained to yield the amount of BST necessary for widespread use.

Practical benefits from the natural functions of these hormones are possible today because rDNA technology allows transfer of the gene for the specific somatotropin (bovine, porcine, etc.) into bacteria that in turn can produce large quantities of somatotropin, in pure form, at a relatively low cost. The rDNA process is similar to that approved by FDA for production of human insulin and human somatotropin.

Many studies have now been conducted using BST in dairy cattle. While under experimental conditions milk production has been increased as much as 40 percent for short periods, under commercial conditions BST is expected to increase annual milk production per cow about 10 to 15 percent and feed efficiency about 5 to 10 percent. Common milk constituents and their proportions are unaffected by the use of BST. Since BST is naturally produced by cows, it is a normal constituent of milk. The level of BST in milk is not changed when cows are treated with BST.

The route and frequency of administration required to make supplemental BST effective present difficulties in making the treatment practical for general use in the industry. Since BST is a protein, it cannot be fed because digestive enzymes would break it down, destroying its activity. Most research trials have been done using daily injections. Several different prolonged-release formulations have been field tested, and it appears that daily injections will not be necessary.

Photosynthesis, the complex process by which green plants convert sunlight to energy, possibly could be made more efficient, thereby allowing plants to grow faster.

The effects of porcine somatotropin (PST) in swine are especially exciting, since the efficiency of production is increased while the amount of intra- and inter-muscular fat produced is dramatically decreased. Research trials have shown improvements between 10 and 45 percent in rate of gain and 15 to 35 percent in feed efficiency. Backfat reductions from 15 to 70 percent have been achieved with corresponding increases in loin eye area of 10 to 50 percent.

Altering ruminal microorganisms for improved performance and feed use by ruminants is an interesting area for the application of biotechnology. Ruminal microorganisms enable ruminant animals to utilize ligno-cellulose in forages; to use significant amounts of nonprotein nitrogen, thus saving protein feedstuffs for other species; and to produce some of their own vitamins.

Several thrusts are of interest for engineering improvements into ruminal microorganisms: controlling degradation of cellulose and nitrogen in feedstuffs; enhancing production of beneficial nutrients such as vitamins and amino acids; controlling the end products of fermentation; and redirecting carbon dioxide and methane to the useful product acetate, rather than the waste product methane. So far, the development of gene transfer systems for movement of cloned genes between ruminal bacteria and *E. coli* has not been very successful.

Bovine Somatotropin

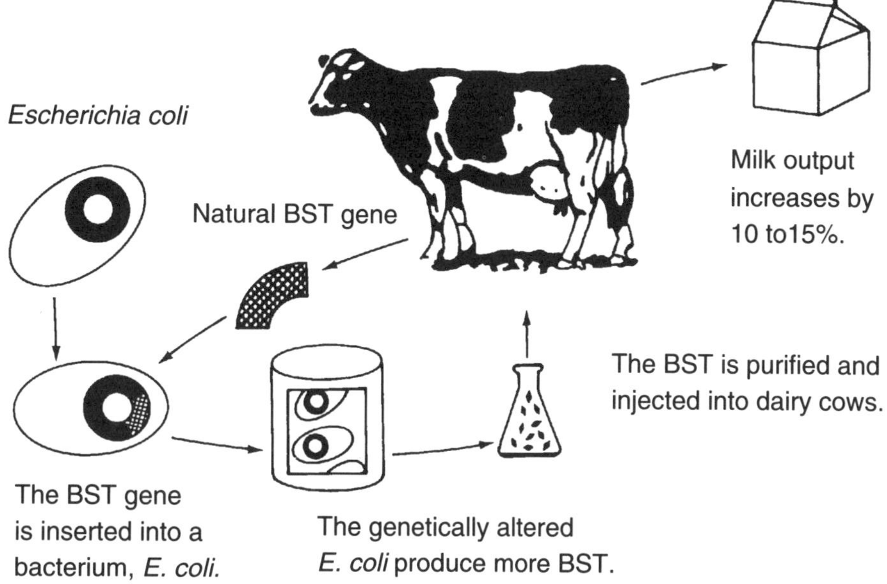

Reproduction

Selective breeding, coupled with artificial insemination and the use of frozen semen to multiply the number of pregnancies possible from one collection of semen, can be considered the first development in animal biotechnology. For several years, newer biotechnologies in the form of embryo manipulations have been utilized, especially in cattle. Embryo manipulation, which includes embryo collection and transfer, or "artificial embryonation," involves the inducement of superovulation in the donor cow (or sheep or goat), which is then artificially inseminated. The resulting embryos are removed for implantation in surrogate females. Before transfer, the embryos can be split, fused (with embryos of other animals), sexed, or frozen. It is now possible to place fertilized embryos from genetically superior parents into other recipient cows, thereby allowing the superior animals to produce more calves. With current technology, 20 to 30 genetic offspring can be obtained from one valuable donor cow per year.

The splitting or microsurgical division of embryos into two, three, or four groups of cells has been accomplished, producing identical twins, triplets, or even quadruplets. The success rate is excellent, and thousands of calves have been produced from split embryos. Currently, the major uses of these technologies are to produce several animals from individuals with especially valuable genetic backgrounds and to provide genetically identical subjects to reduce the variability in research on animal nutrition and physiology.

The ability to determine the sexual identity of sperm and/or embryos would have benefits in animal production. Commercial dairy production implies an obvious preference for milk-producing females. In swine, females are preferred, since they grow faster, whereas in beef cattle and sheep, the male grows faster and thus would be preferred for meat production. Another advantage of embryo sex determination would be to speed up genetic improvement, since desirable traits are often associated with one or the other parent. Research on techniques for sexing sperm/embryos is continuing because the sexing cannot yet be achieved accurately on a consistent basis.

Animal Health

Animal health is essential to the profitability of livestock production, and much improvement is needed in the diagnosis, prevention, and control of animal diseases. Most new diagnostic aids derived from biotechnology will employ one of three technologies: monoclonal antibodies, DNA probes, and immunosorbent assays. Recombinant DNA can be used to make "probes" to diagnose disease. A DNA probe is a molecule that has been labelled with a radioactive isotope, dye, or enzyme. It is used to locate a particular nucleotide sequence or gene on a DNA molecule and thus identify the pathogens responsible for various infectious diseases. Such probes have potential for fast, accurate diagnosis of disease.

"Monoclonal antibodies" (MAs) is the term applied to a group of identical antibody molecules which recognize only the specific antigens displayed by the infectious agent. They can be used to diagnose disease, monitor drug efficacy, and develop therapeutic treatments and vaccines to immunize against diseases. Currently, diseases such as scours, pseudorabies, bluetongue, transmissible gastroenteritis, and several major diseases in poultry can be diagnosed. Precision of diagnosis is excellent. Monoclonal antibodies currently are available as therapeutic treatments against both calf and pig scours. Other examples of uses include detection of estrus and pregnancy in cows and of aflatoxin in feed.

Safer and more effective vaccines are being developed with biotechnology. Until recently, the primary source of antibodies used for immunization and other purposes was blood serum from many animal species. However, such serum also contains antibodies to hundreds of other substances, and each antibody type was limited in quantity. Some rDNA vaccines known as "subunit vaccines" stimulate immunity using only a portion (or subunit) of the virus or bacterium which has been isolated from a pathogenic organism. Another type of vaccine already is on the market for swine pseudorabies, a frequently fatal herpes disease. This vaccine has the actual disease-causing genes removed from the live virus, which can then be injected. Vaccines for scours, foot and mouth disease, and rinderpest also have been developed through biotechnology.

Animal Genetics and Transgenic Animals

Manipulation of specific genes in livestock is now possible because of biotechnology. Modifying the genotype of animals to improve disease resistance, stress tolerance, and growth performance is possible. As was noted earlier, methods to transfer genes and to control their expression are key problems in animals, as well as in plants. The desired gene(s) must be identified, isolated, and cloned. Usually a DNA construct (a gene with a regulatory sequence) must be developed, placed in a vector, and cloned in bacteria. The gene can then be cut out of the bacteria and made ready for transfer to the recipient animal.

By injecting genes into one-cell embryos, it is possible to produce

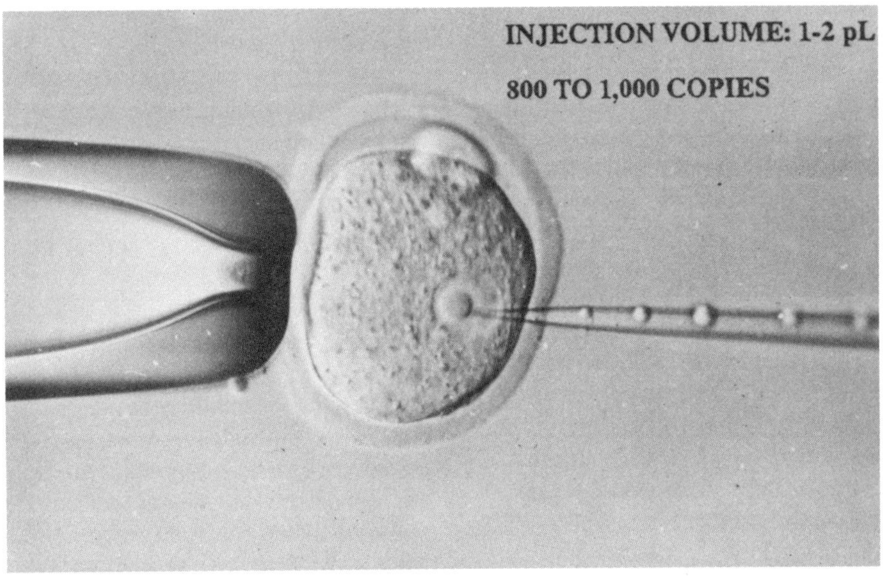

Microinjection of fish eggs makes it possible to add foreign genes into the genetic makeup of the fish that will develop from the eggs.

animals that have the new genes in every cell. Animals produced with genes added in this way are called "transgenic animals." The production of transgenic animals has obvious advantages over frequent injections with compounds such as somatotropin produced by biotechnology into nontransgenic animals. The genes to be injected could come from animals of the same or a different species. Even this technique mimics nature, because genes are constantly being moved around within, and occasionally between, species by some types of virus infection.

The first transgenic mouse was reported in December 1982 and received much attention in the media. These classic experiments in mice showed a doubling of the mature size in transgenic mice which contained extra copies of genes for either rat or human growth hormone. This larger strain of mice has maintained its larger size through successive generations. Since then, medical scientists have successfully implanted human genes into mice to create several animal models of human disease for research on those diseases. More recently, gene transfer technology has been applied to commercially important animal species, including swine, cattle, sheep, rabbits, chickens, and fish.

Three ways of approaching gene transfer in animals are being pursued. The simplest way is to inject a gene into a one-cell embryo (into the male pronucleus of a fertilized egg) and then implant the embryo into a surrogate female, in the case of mammals. When the conditions are right, the new gene is incorporated into the DNA (genome) of the target embryo so that, as the cell divides, the new gene is in the DNA of every cell.

A second approach is to incorporate the gene into a type of virus known as a "retrovirus" that has been modified so it cannot reproduce itself after entering the cell. This method has the advantage that the disarmed retrovirus always delivers just one gene that is always intact.

A third method is to add genes to undifferentiated embryonic cells, which are then placed in other embryos. However, the difficulty of keeping the embryonic cells of farm animals from differentiating into various specialized cells presents a barrier for use of this technique with these species.

Genes for disease resistance, more efficient growth, and enhanced protein synthesis would be likely candidates for transfer. Among the strains of commercially useful transgenic farm animals likely to emerge in the next few years are leaner pigs, poultry resistant to avian influenza, sheep with wool that is easier to wash, and goats and dairy cattle that produce valuable pharmaceuticals in their milk. Efforts are underway to genetically engineer animals to produce more of their own somatotropin. While this potential has been demonstrated in swine, cattle, sheep, and goats, several undesirable side effects have been noted.

More needs to be learned about the control of gene expression before the production of transgenic animals can be perfected. The ability to produce animals exhibiting only the beneficial effects of somatotropin by a transgenic approach may require better control of expression, a specific starting genetic background, and/or a modified husbandry regimen. For broader application of transgenic possibilities in farm animals, much additional research also is needed on gene mapping (as in plants) so that the tools of rDNA can be fully utilized. Most of the benefits from direct genetic engineering of farm animals probably are still some years away. Genetically engineered fish may be the first such development to be commercialized.

Gene mapping, identifying the specific location of important genes, is of great importance in animals and humans, although it has been slower coming than gene mapping of plants (discussed earlier). Much attention has been given lately to the creation of a major national effort to map the human genome. One reason that animal gene mapping has been so late in coming is because gene mapping is much more difficult in animals than in plants. It now appears that in terms of gene location, cattle may be much more similar to humans than many had suspected. Thus, the high resolution genetic maps of the human chromosomes now being developed can be a useful guide for animal genes as well. A significant commitment of public funds will be necessary to launch a meaningful effort in mapping the animal genome.

Another interesting application of gene transfer technology is using goats, sheep, or cattle as bioreactors. Successful demonstrations to date are few, but sheep and goats have been genetically engineered to produce milk containing pharmaceuticals such as tissue plasminogen activator (tPA), which dissolves blood clots after heart attacks; erythropoietin (EPO), which regulates the production of red blood cells; and human blood-clotting Factor IX, which hemophiliacs use.

The Evolution of Technology Assessment

The advances in agricultural biotechnology discussed in the preceding section offer substantial benefits to society. Yet, careful scientific assessment is necessary to insure that these products are safe and efficacious. Also, the potential socioeconomic impacts of the products must be addressed to help society make appropriate choices and adjustments regarding adoption of the products and the new technologies.

For many years, the U.S. government has required federal approval of new food and health products (human and animal) before they could be commercially sold. Later, registration of pesticides and other toxic substances was mandated. Various government agencies have authority for regulating different substances. In the United States, regulatory agencies such as the FDA or Environmental Protection Agency (EPA) base their evaluations on scientific criteria, including product efficacy, product safety, and environmental implications. Social and economic effects are generally determined by market forces and occasionally by legislative action. However, with the emergence of agricultural biotechnology, there are pressures in the United States and in the European Community for product review to include both technical and socioeconomic assessment before a product is approved for use.

Scientific Criteria for Product Evaluations

- *safety*
- *efficacy*
- *environmental implications*

The economic and broader societal impacts of current and emerging agricultural technologies are often analyzed by social scientists. Prior to World War II, the socioeconomic assessment of agricultural research and technology transfer was primarily a process of descriptive documentation. Extension agents, farm journalists, and others noted that farmers were substituting machines for horses and manual labor, making genetic improvements in their farm animals, and adopting improved crop varieties, most notably hybrid corn. Better farm management was increasing farm productivity as demonstrated by higher crop yields and improved livestock feed efficiency.

Since World War II, agricultural economists and rural sociologists have greatly expanded their conceptual basis for and empirical assessment of agricultural technology. Improvements in databases and computers have made advanced statistical analysis possible. Most studies have examined the private and societal returns to agricultural research and extension investments, and the distribution of benefits and costs. Others have focused on technology adoption patterns.

Until recently, most of the socioeconomic assessment of agricultural technologies was conducted *ex post*, i.e., looking back at technologies that farmers had already adopted. With the development of agricultural biotechnology, society is demanding *ex ante* research that requires the assessment of new technologies before they are fully developed, approved by government regulatory agencies, or adopted by farmers. *Ex ante* research forces analysts to make judgments about uncertain future events such as adoption rates, productivity changes, farmer and consumer reactions, and costs of adopting the technology.

Several recent socioeconomic studies of agricultural biotechnology have employed an *ex ante* approach. These include: BST adoption by dairy farmers (Fallert, et al.); the adoption of PST in swine production (Lemieux and Wohlgenant); and the general use of growth hormones in livestock production (Kuchler and McClelland).

There has been relatively little economic assessment of biotechnology applications in crop production and food processing. This may be attributed to data limitations and to the fact that biotechnology applications in crop production are not as well developed nor as well understood at this time as are applications in livestock and dairy production. Also, there has been greater public concern and controversy associated with imminent farmer adoption of certain livestock products of biotechnology, while public concern for biotechnology applications in the crop area have mainly focused on field testing up to this point in time.

Society is increasingly concerned about the social dimensions of expected technological impacts. Historically, technology approval has been based on evidence that the technology is efficacious and safe to users, any livestock involved, consumers, and the environment. Today, many seek information concerning the distribution of economic effects, e.g., small versus large farmers, poor versus rich consumers, local versus multinational input suppliers, and one geographic region versus another.

The public wants to be included in the assessment process. Many do not want to leave the approval of technology to scientists alone. Hence, there is a growing national debate over agricultural biotechnology that involves diverse groups. And much of this debate is centering on the social and economic ramifications of agricultural biotechnology. Public perceptions and attitudes towards agricultural biotechnology may delay, or even prevent, the release or adoption of new biotechnologies. The current BST controversy is illustrative of the new role the public is assuming in technology assessment. Consequently, objective *ex ante* socioeconomic assessment of agricultural biotechnologies is needed, and this information must be shared with the public.

This chapter has provided an overview of the science of biotechnology, current and potential applications in agriculture, and the social and economic issues surrounding agricultural biotechnology. Subsequent chapters in this book explore these issues more thoroughly, drawing from the knowledge and experience of many experts, including agricultural economists, rural sociologists, animal scientists, molecular biologists, food scientists, biochemists, agronomists, and veterinarians. The contents of this book should enable readers to make more informed choices concerning the development and application of agricultural biotechnology.

References and Additional Readings

Baumgardt, B.R. 1988. Biotechnology and the animal sciences. In *W.B. Lacy and L. Busch (eds.) Biotechnology and agricultural cooperatives. Opportunities and challenges. Kentucky Agricultural Experiment Station. Lexington, KY.*

Council for Agricultural Science and Technology. 1986. Genetic engineering in food and agriculture. CAST Report No. 110. (Available from CAST Headquarters Office, 137 Lynn Avenue, Ames, Iowa 50010.)

Fallert, Richard, T. McGuckin, C. Betts, and G. Bruner. 1987. bST and the dairy industry: A national, regional, and farm-level analysis. U.S. Dept. of Agric., Economic Research Service, Agricultural Economic Report No. 579.

Gasser, Charles S. and Robert T. Fraley. 1989. Genetically engineering plants for crop improvement. Science 244:1293-1299.

Griliches, Z. 1958. Research costs and social returns: Hybrid corn and related innovations. Journal of Political Economy 66:419-431.

Kuchler, Fred and John McClelland. 1989. Issues raised by new agricultural technologies: Livestock growth hormones. U.S. Dept. of Agric., Economic Research Service, Agricultural Economic Report No. 608.

Lemieux, Catherine M. and Michael K. Wohlgenant. 1989. Ex ante evaluation of the economic impact of agricultural biotechnology: The case of porcine somatotropin. American Journal of Agricultural Economics 71:903-914

Lindow, S. E., N. J. Panopoulus, and B. McFarland. 1989. Genetic engineering of bacteria from managed and natural habitats. Science 244:1300-1307.

National Research Council. 1987. Agricultural biotechnology, strategies for national competitiveness. Committee on a National Strategy for Biotechnology in Agriculture. Board on Agriculture. National Academy Press, Washington, DC.

Purchase, H. Graham and David R. MacKenzie, eds. 1990. Agricultural biotechnology: Introduction to field testing. U.S. Dept. of Agric., Office of Agricultural Biotechnology, Washington, DC.

Riepe, Jean Rosscup and Marshall A. Martin. 1989. Biotechnology: Implications for animal agriculture. Journal of the American Society of Farm Managers and Rural Appraisers, Vol. 53, No. 2, p. 47-52.

Watson, James D., John Tooze, and David T. Kurtz. 1983. Recombinant DNA, a short course. Scientific American Books, W. H. Freeman and Company, New York, NY.

CHAPTER TWO

Biotechnology is only one technology among many which are and will be sources of change in American agriculture and rural communities. While biotechnology adoption will not seal the fate of U.S. agriculture, it is likely to be a significant driving factor in the evolution of future agricultural production systems. Specific developments may have substantial transforming effects on the production and input supply sectors while offering needed productivity gains and environmental benefits. Other biotechnology innovations will have only subtle or minimal impacts on production systems but will still benefit producers, processors, and/or consumers.

Biotechnology is already affecting the input supply sector of U.S. agriculture and will increasingly affect the production sector over time. Increased concentration and consolidation have already occurred in the seed and agricultural chemical industries. Although biotechnology should have a supply-increasing effect in the production sector, impacts among individual production units will differ.

Whether the advent of biotechnology will hasten the trend toward fewer, larger farms will depend in part on how adoption affects the cost structure of farms. Innovations from biotechnology are not likely to be inherently detrimental to the family-farm structure of American agriculture. However, it may be that for smaller scale production units to effectively benefit from the adoption of these new technologies, support will be required in the form of targeted research and development, market access, and tailored extension information services. Choices made by decision makers in society about which or what type of biotechnology products to develop and about the informational network to support product adoption will influence how biotechnology will affect the agricultural sector and rural communities.

2 Emerging Biotechnologies: Impact on Producers, Related Businesses, and Rural Communities

W. Burt Sundquist, University of Minnesota
Joseph J. Molnar, Auburn University

Introduction

Social and economic change associated with scientific advance is an integral part of modern life. The emerging biotechnologies present a next major stage in technical change in agriculture (Hess, 1987). In this chapter we assess potential changes in agriculture, forestry, and aquaculture and discuss the consequences these shifts might have for related businesses and rural communities.

Biotechnologies are generally expected to reduce the cost of food and improve the quality and quantity of outputs from these industries. Much of the concern over biotechnologies is directed to their potential to accelerate trends toward fewer and larger production units. They also may increase the influence of large corporations and other off-farm actors on the decision making and fate of farmers and local community residents.

The evolution of technology in agriculture in recent decades and its key impacts are noted briefly in the discussion which follows. We also address the future structures of both the production and the supply-service sectors, as well as the nature of public perceptions which may influence development and adoption of biotechnology products. We begin with a brief perspective on the changing technical, economic, and structural context for food and fiber production in the U.S.

Biotechnology and Socioeconomic Change in U.S. Agriculture

Historical Perspective

In the mid-1930s, U.S. farm numbers peaked at about 6.8 million, and total average annual farm workers numbered 12.7 million. Substantial mechanical technology was already in place on U.S. farms, but its rate of adoption was slowed by the disastrously low farm prices and incomes of the Great Depression.

World War II brought much higher farm income, and, with the end of the war, raw materials became available for the farm machinery industry to supply the pent-up demand of farmers for modern mechanical technology. As a result, between 1947 and 1979, mechanical power and machinery inputs doubled on U.S. farms.

The chemical nitrogen manufacturing industry, armed with technology developed for World War II munitions production, shifted much of its capacity to supplying the emerging research-generated demand for nitrogen for farm fertilizers. Cheaper fertilizer, together with the subsequent development and use of chemical pesticides, increased the use of chemical inputs more than eightfold between 1947 and 1979. Combined with new crop varieties and changes in other cultural practices, increased fertilizer use pushed up crop yields dramatically. For example, U.S. corn yields averaged only 24 bushels in 1935 but exceeded 100 bushels per harvested acre by the late 1970s.

On the animal side, research-based improvements in breeding and nutrition were combined with effective disease-control technologies and improved management practices. As a result, U.S. production of livestock and livestock products increased by 60 percent between 1947 and 1979, and rapid productivity gains, particularly for broilers, fueled dramatic output increases of 160 percent for the poultry and egg sector.

Social and economic change associated with scientific advance is an integral part of modern life.

As a result, U.S. agriculture was transformed over a few decades from a largely resource-based sector (mainly land and labor) to one with a high degree of dependence on inputs purchased from off the farm. Moreover, farm production became much more specialized along commodity lines. As a result, the 1987 U.S. Census of Agriculture reported fewer than 2.1 million farms and 3 million total average annual farm workers. Moreover, of the total of $136 billion in annual product sales, more than 75 percent came from the fewer than 300,000 farms with annual product sales of $100,000 or more. For the reasons cited above, any future changes in farm numbers and farm employment resulting from biotechnology or other sources can never be as great as the changes of the last several decades.

Despite the dramatic changes in the structure of farming illustrated here, in 1987 almost 87 percent of all U.S. farms were still operated as sole proprietorships. Partnerships accounted for another 9.6 percent, and family corporations about 3 percent. Other-than-family corporations totaled only about 0.3 percent of all farms, but they were very important in the total production of some agricultural commodities. On balance, family-scale units continue to dominate production in most types of U.S. farming, but many are large, specialized units with heavy reliance on production inputs purchased from off the farm (Office of Technology Assessment, 1985).

Rapid internationalization of financial markets in the early 1970s coincided with a shortage of worldwide food supplies and much expanded trade in farm commodities. U.S. farmers responded by adopting available technology and expanding output dramatically. Subsequently, excess supplies resulted in price declines for major commodities. But excess supplies masked the fact that by the early 1980s the rates of productivity gains from mechanical and chemical technologies had slowed appreciably from those of the preceding decades. Thus, with mounting global food needs, the emerging biotechnologies could well be the major source of productivity gains necessary to meet these food demands and the challenges of growing global competition.

Mechanisms Underlying Biotechnology Impacts

Biotechnology products represent the next stage in the application of science and technology to U.S. agriculture. But there are differences. Most previous technologies were capital intensive and substituted heavily for labor and land. Many emerging biotechnologies will substitute for conventional purchased inputs. An example is the substitution of biopesticides for some chemical pesticides in plant insect control. Another is the substitution of biotechnology-improved animal disease vaccines for existing ones.

Some biotechnologies will complement existing technologies. An example is the genetic transformation (engineering) of plants to incorporate desired traits. In the use of this type of biotechnology, conventional plant breeding will still be required for incorporation of biotechnology-induced traits into commercial lines, for continued plant improvement selection, and for seed multiplication. Also, for the foreseeable future at least, chemical fertilizers will remain important in crop production. A key question is whether biotechnology applications will, on balance, be complementary to "capital-intensive" or to "low-input" farming systems. The latter type of system is frequently considered to be the more supportive

While most previous technologies were capital intensive and substituted heavily for land and labor, many emerging biotechnologies will simply substitute for conventional purchased inputs.

of family-scale farming systems (Hassebrook and Hegyes, 1989).

The happiest consequences for farmers stem from biotechnology innovations that lower their production costs per unit of output, help them cope with environmental adversities, and/or allow them to enter new or expanded product markets. Though suppliers of cost-reducing technologies will not pass all savings on to the farmer, competition among suppliers will generally ensure some benefits at the farmer (producer) level.

Biotechnology may also enhance the performance of existing inputs by improving specificity, shelf life, or other utility dimensions that benefit farmers. Improvements in the quality and performance of a number of pharmaceuticals and microbial agents are expected to offer clear benefits for both input suppliers and early-adopting farmers.

As with past technologies, successful adoption of specific biotechnology innovations will require additional profits for some, at least the early adopters. Moreover, increased profits, as in the past, will result mainly from reductions in real production costs per unit of output. If these cost reductions are realized by farmers and other producers (and not captured mainly as profits by the manufacturers and distributors of the innovations), they will increase productivity and the competitive position of U.S. producers. But, as before, there will be both winners and losers.

Differential Effects of Innovations

Biotechnology, like most past technology innovations, is expected to be supply increasing in the aggregate (Reilly, 1987). Nevertheless, the implications of new technologies can be quite different for each individual farm, given the source and nature of the product, as well as the economic, human-capital, and life-cycle situation of the farm operator.

Some potentially dramatic improvements in productivity have been regarded as a threat to particular categories of farmers and certain industries. Following the analogy of the agricultural treadmill, late adopters will be faced with lower product prices. Early-bird farmers have already cut their costs, enjoyed wider profits in their period of initial adoption, and are ready to respond to the next wave of technical innovation (Cochrane, 1979). Increased supplies are generally associated with lower prices. As a consequence, nonadopters often have higher costs while facing lower prices for their production.

Some biotechnology products will be readily substituted for the old seeds, pesticides, vaccines, or other inputs previously used in the production process. But other biotechnology products will require enhanced management and technical understanding if their full benefits are to be realized.

More generally, whether biotechnology innovations will significantly accelerate the trend toward fewer and larger units depends on how they affect the cost structure of typical production firms. If the technologies require substantial capital outlays so that fixed costs of production rise appreciably, smaller firms will be placed at a competitive disadvantage relative to larger, better capitalized firms because total per unit production costs after adoption of the technology are higher for a small firm.

Conversely, if the new technology is relatively inexpensive, affecting only variable costs, significant bias in favor of larger firms will not exist. Nevertheless, even low-cost, highly divisible technologies can adversely affect small producers to the extent that they contain a fixed-cost element emanating from the need to acquire and assimilate information about the technology for its effective use.

New classes of input innovations stemming from biotechnology present a mixed picture for the agricultural producer. Such products involve a discrete shift in the production process and may require implementation of a qualitatively different technology package. Bovine growth hormone, nitrogen-fixing soil inoculants, or frost-inhibiting bacterial plant treatments are examples.

Such new classes of inputs will sometimes require increased human as well as monetary capital for successful implementation. Where complex situational assessments and technical expertise are required to obtain full benefits of an innovation, less educated farmers with limited capital resources face additional barriers to participation in the promise of a new technology. Better capitalized farmers can hire the necessary expertise if they do not already possess it. Thus, new classes of biotechnology innovation are most apt to widen differences between large and small farms (Teigen, 1986). An exception could occur where the additional management requirements are tied closely to the labor input. In this case, owner-operator farmers can be expected to outperform unskilled hired workers in the use of new technologies.

Differences in availability, timing, and supportive circumstances for new innovations can widen existing inequalities among farmers and create new ones. Specific innovations may have certain agroclimatic performance traits that create regional disparities in effectiveness. Also, some farmers are in locales better served by markets and services supporting the use of the innovation. As a result, farmers in marginal production areas for a particular commodity may have less contact with other producers experienced with the innovation and, thus, delay adoption. In turn, national or regional suppliers may delay making an innovative product available in those areas where few producers exist because the "thin market" does not justify promotion and distribution costs. Local suppliers may exhibit a similar reluctance to stock or promote a new product.

Some existing production inputs will become obsolete, and their manufacturers and distributors will lose markets. A number of producers who either do not adopt cost-reducing biotechnologies or who adopt but do not manage them well will be disadvantaged in their competitive position, some seriously so. And whether biotechnologies are adopted or not, many agriculturally based rural communities will continue the ongoing process of shrinkage and consolidation as producers and local supply and marketing firms continue to decline in numbers.

Commercializing Biotechnology Innovations

The science that gave rise to biotechnology is only beginning to be translated into commercial products for farm use. Private-sector investment plays a vital role in commercialization of agricultural biotechnologies. While agricultural biotechnology sales in the U.S. probably totaled about $50 million in 1989, they are projected to possibly go over $2 billion by the year 2000 (Genetic Engineering News, 1989).

It should not be concluded that forthcoming biotechnology applications are predetermined. Priorities for biotechnology research, particularly for publicly funded research, can still be established or modified to help achieve desired structural and environmental goals. Also, health and environmental risks must be considered for each new item. The regulatory process will likely prevent approval of some, but experience and broader public understanding will increase the routine acceptability of large classes of new innovations.

A number of factors affect the rate and extent to which individual biotechnology products or processes will be commercialized. Some of these are mainly supply-side considerations (by technology developers and distributors), others are mainly demand-side (by users and public interest groups), and some are both. It is not our objective to predict which specific products or processes will be commercialized but only to identify the factors affecting the likelihood that a biotechnology product will reach the farmer.

On the supply side, what technologies are made available to the farmer or other users is a function of: a) costs of product development and marketing, b) expected market volume, c) expected time and costs to meet regulatory requirements for products and processes, d) facility for small-scale

Whether biotechnologies are adopted or not, many agriculturally based rural communities will continue to undergo the process of shrinkage and consolidation.

Figure 1. *Aggregate Adoption Profile for Technologies.*

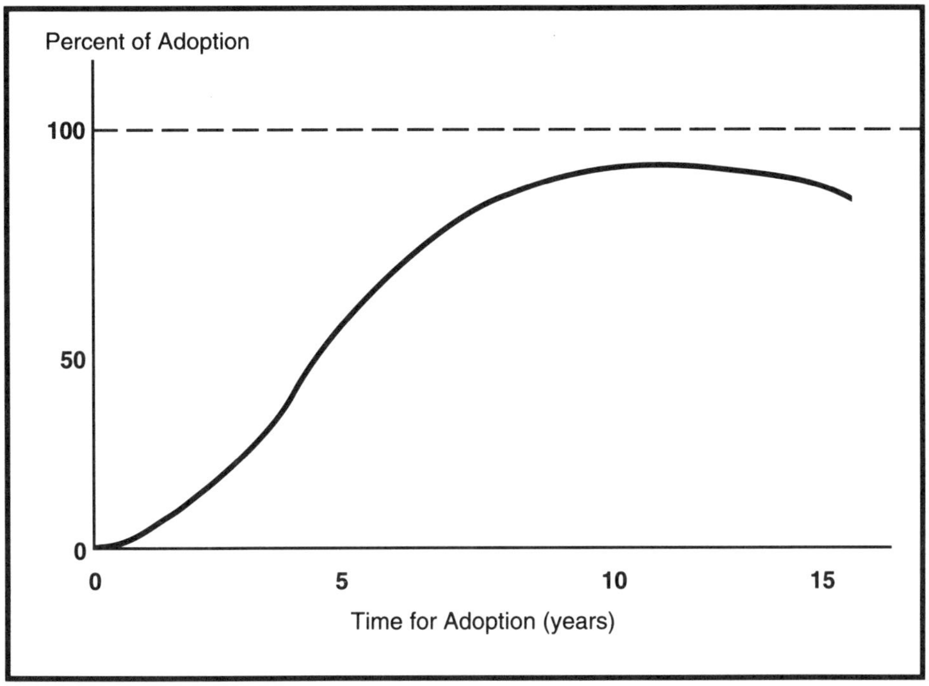

product testing, and e) expected profitability for suppliers (both absolutely and relative to other investment alternatives).

On the demand side, factors affecting what farmers implement in their operations include: a) costs of biotechnology products, b) their expected effects on yields and/or product quality, c) perceived safety in use, d) degree of technical sophistication required for effective use, e) the amount and sources of information available to potential users, and f) bottom-line profitability in use. Also of importance will be expected consumer reaction (demand response) for commodities produced with biotechnologies.

In addition, both supply and demand for biotechnology innovations will be affected by a) national and international economic and regulatory policies and outlook, b) commodity-specific policies and outlook, and c) public attitudes toward product safety and desirability. The latter include human health and safety concerns, environmental concerns, and broad ethical and social policy considerations. Compared to the commercialization and use of prior agricultural production technologies, consumers, environmentalists, animal-welfare advocates, and other public interest groups are taking a much more active role in the public policy discussions on biotechnology products and processes. So are producers. And this broadened participation can be expected to become even greater as more biotechnology products approach the stage of marketability.

Adopting Biotechnology Products

The rate and extent of innovation adoption is important to both the manufacturers and the users of biotechnology products. For example, the size of market and the profit potential for suppliers depend on adoption rates. And adoption rates will affect aggregate producer commodity supply response (and prices), as well as the competitive position of individual firms (Teigen, 1986).

As in the case of many earlier production technologies, not all producers will adopt specific biotechnology innovations. And for those who do, adoption will not occur simultaneously. Historically, a number of factors have influenced technology adoptions. These include, as mentioned earlier, cost of adoption, the amount and availability of information for the technology, the technology's expected profitability in use, and how complementary the technology is within the existing production system. But they also include such factors as firm size and location; the age, education, and experience of producers; and whether or not they are aggressive innovators of new technologies.

Historically, technology adoption rates have aggregated to an S-shaped curve of the general type shown in Figure 1, with the speed of adoption depending on investment, learning requirements, and on profitability in use. In general, the earliest adopters in agriculture have tended to be young to middle-aged operators who manage larger commercial units (Rogers, 1982).

Preliminary indications are that adoption patterns will be similar for biotechnology innovations as for their predecessor technologies. But there is strong reason to believe that adoption of some biotechnology products will be strongly resisted by some producers for a variety of reasons. Technological change has not always been welcome in American agriculture, and some biotechnologies may be identified as particularly threatening in some quarters. Thus, the adoption process will be moderated by the social standing of biotechnology innovations that may affect producer attitudes, consumer acceptance, regulatory processes, and, in some cases, all segments of the food system.

Significant producer or consumer resistance has not been voiced for the potential adoption of most economically profitable biotechnologies. But, as mentioned earlier, there is concern by some that biotechnology applications should be complementary to "low-input" rather than "high-input, capital-intensive" production systems on the grounds that the former is more supportive of owner-operator, family-scale farming. Also, the continuing issue of the release of modified or even existing organisms into the environment may eventually influence adoption rates for some crop-related biotechnologies.

Expected Impacts on Crops, Livestock, Forestry, and Aquaculture

Over the next several decades, emerging biotechnologies will have major and pervasive impacts on U.S. producers of farm and forestry products. But many of these changes will occur on a gradual basis. Regulatory procedures typically require five to seven years between the time of initial product identification to its commercial availability. Even then, initial market entry will generally be on a pilot basis. This suggests that most biotechnology products which will be marketed within the next five years or so must already be well along in the development process.

No one can fully anticipate the impacts of biotechnology on the agricultural, forestry, and aquaculture industries. But, to the extent that some insightful perspectives can be gained, it can probably best be done by viewing separately each of the several major affected sectors.

Crops

As suggested earlier, agricultural production has become more specialized over time along commodity lines, although most farmers still produce multiple commodities. Within the crops sector, it is important to contrast large-acreage volume crops, such as wheat, corn, cotton, and soybeans, with the smaller acreage specialty crops, such as tomatoes, potatoes, and onions. Several important distinctions exist between these categories of crops.

First, the large-acreage crops are not highly coordinated or integrated across their input, production, and marketing stages, whereas some small-acreage specialty crops are. Second, the potential size of market for new technologies such as biopesticides is much greater for the large-acreage crops than for the specialty crops. And potential "size-of-market" will be an important driving force for the development of future technical innovations. Third, biotechnology processes are already available to alter the harvestable component of some specialty crops. While there are some special market-use exceptions, the latter developments appear further away for the major food and feedgrain crops and for cotton.

New biopesticides, quick pesticide residue tests, and disease- and pest-resistant crops are even now entering or approaching entry into the marketplace. Selected biopesticides are now at the product-testing stage. Quick, on-site immunoassay tests can be used to detect residues on crops and allow for on-site checks by the grower to see if there is a contamination problem. Progress has also been made in increasing the biological nitrogen-fixation capability of some legume crops. Biological nitrogen fixation in cereal crops, however, remains a seemingly distant objective.

Although chemicals will likely continue to be the chief form of crop pest control for the next 10 to 15 years, innovations in this area are being induced by deep and widespread public concerns over the relative safety and long-term impacts of current agricultural pest control agents.

Biopesticides hold the dual promise of rapid and well-understood degradation, as well as overcoming pest resistance to conventional chemicals, an increasing problem for vegetable crops.

Other developments relate to genetically improved insect and viral resistance in cotton, tomatoes, potatoes, and other crops. Such advances are particularly important in cotton, which has one of the heaviest chemical application rates of any agricultural crop. To the extent that pest resistance can be incorporated into the plant and not applied by aerial spraying or otherwise, cumulative environmental impacts can be mitigated. The pressure to identify comparatively safer biotechnology alternatives to conventional chemicals will only grow as epidemiological data, other scientific evidence, and public attention continue to focus on the safety and environmental sustainability of the nation's food and fiber system.

Although a number of microbial agents useful for crop production have been available and in use for many years, the release and use of some additional organisms could create new regulatory burdens for farm operators. Producers of some commodities already have extensive experience with organisms as production inputs, yet the potential environmental impact of transgenic organisms is a concern for some.

Concerns about these organisms relate more to the unknown consequences of genetically engineered augmentations to the organisms, however, and not generally to the impacts on farmers or farm industries (Tiedje et al., 1989). But, to the extent to which the use of selected microbial or other biotechnology-based agricultural inputs are associated with additional permitting, registration, or notification processes, at least some small-scale farm operations will probably forgo utilization of the technologies.

The Large-Acreage Volume Crops

As already mentioned, such biotechnology applications as selected herbicide resistance and biopesticides are near-term realities. Biotechnology applications that are embodied in purchased seeds or chemicals can be used and managed by virtually all farmers. Unlike previous mechanical technologies, most biotechnologies will not, in themselves, generate significant size economies. Moreover, there appears to be little incentive for expanded vertical integration into crop production by those firms supplying seed and chemical inputs. Unanswered questions are the extent to which (1) new seed-herbicide technology packages may overcome existing constraints (such as weed control) for high-input (capital intensive) crop production systems and/or (2) costs of new inputs may be discounted to large-volume purchasers. It is through these mechanisms that there exists the potential for farm-size-increasing effects from the new seed-chemical technologies.

In the longer term, field crops are expected to be affected significantly by biotechnologies which are yield increasing. Such developments, when they occur, will likely have supply-increasing and price-dampening effects which are expected to adversely affect the staying power of higher cost producers. But the commercialization of significant yield-increasing biotechnologies must, in any event, generally await the technical capability to transfer multiple genes. Currently this capacity is not yet even at the experimental stage.

Biotechnology innovations and impacts are likely to differ between large-acreage crops, such as wheat and corn, and small-acreage specialty crops, such as potatoes, onions, and the tomatoes pictured here.

Farmers are already being affected by innovations that shift demand among commodities by allowing formerly distinct feedstocks to substitute for one another. Currently, soft drink and other food manufacturers are replacing sugar with high fructose corn syrup. The world sugar market is highly depressed, and many Third World nations are scrambling for alternative export crops. Moreover, in the absence of large public (protection) subsidies for U.S. sugar producers, the economic status of these producers would already be seriously impaired. Similarly, advances in enzyme technology may reduce the utilization barriers among vegetable oils, thus increasing the market competition for U.S. oilseed producers from palm oil and other potential substitutes.

With respect to environmental implications, there is the strong possibility through such products as biopesticides, and perhaps ultimately with biological nitrogen-fixation capacity for cereal crops, to reduce the reliance on chemicals in crop farming. In the near future, however, there appears little prospect for biotechnologies which will substitute significantly for the use of chemical fertilizers.

Existing Specialty Crops

As suggested earlier, selected biotechnologies already have the capacity to modify the harvestable product for such specialty crops as tomatoes. This capability will increase the extent to which processors specify product quality. It will also tighten the vertical coordination linkages between production inputs (including seed) and the production and processing stages for a number of specialty crops.

Although there are no obvious size economy implications associated with most expected biotechnology innovations, the process of vertical coordination and marketing is generally more easily managed with a smaller number of larger producers than with a larger number of smaller producers. Hence, it will probably be even more difficult in the future for small producers to compete in the specialty crops sector except for special market niches. Organic foods, pharmaceutical inputs, and ethnic items (such as ginseng) are examples.

Biotechnologies will be particularly important in vegetable and fruit crops, where product quality, shelf life, and taste are major salient characteristics. Biotechnology tools that allow greater selectivity in specifying the performance traits of different crop varieties will allow more rapid development of desirable cultivars and much more rapid propagation of plant stocks. The markets for tomatoes, lettuce, carrots, and a few other crops are large and relatively focused on a few specific varieties. Improvements here have high potential for rapid and widespread adoption with benefits for growers, plant stock sources, and consumers. Significant price differentials connected to biotechnology-based improvements also can be expected to raise the importance and perhaps the controversy attached to biotechnologies by the average consumer.

Crops with New Market Niches

Biotechnology offers the possibility of new crops and/or modification of existing crops for food, feed, and industrial uses. Examples are those of modifying the seed composition of corn and soybeans.

The industrial use of corn for glucose, dextrose, starch, and alcohol has expanded rapidly, and research is underway to modify the protein, starch, and oil content of the grain. Currently in the U.S., only about 3 percent of the corn grain acreage is planted to special-use hybrids, including white corn for corn meal and grits, waxy corn for use as thickeners in the food industry, and hard yellow corn for snack chips. The other 97 percent is sold, mainly through country elevators, under the broad market classification of No. 2 yellow corn, without measurement of protein, starch, or other quality characteristics. But it is expected that a number of new special-use corn hybrids will be on the market within three to five years.

Before it is profitable for farmers to grow such special-use corn hybrids (or similar specialty-use products), however, effective new marketing channels will need to be developed to permit producers to capture price-premium incentives for their special-use products. Since it will probably be a number of years before most country grain elevators can accommodate the marketing of specialty varieties, direct contracting between processors and growers is expected to play an important role in their market development and growth.

The potential range of modified existing crops and the development of new crops to service specialty market niches are substantial, and the above example for corn suggests the likely general pattern of marketing channels to accommodate such special-use crops. In those cases where specialty market niches are small in volume, the incentives will be substantial for a high degree of vertical integration in production and marketing. Such circumstances will serve to limit the production opportunities for most independent producers but could provide new opportunities for some small producers.

Impact of Specialization

Increased specialization within farm production systems has been a major source of farm productivity growth and increased enterprise size. As a result, livestock enterprises have disappeared from many farms, and only cash crop enterprises remain. Among the consequences have been an intensification of row crop production in some regions and elimination of animal manures and legume crops as sources of nitrogen supplies on many specialized crop farms. The move to shortened crop rotations and, in some cases, to monoculture has stimulated the build-up of some plant pest populations. This has contributed to the intensified use of chemical pesticides.

Labor efficiencies and lower investment costs per unit of output for machinery, equipment, and buildings are among the key economic forces that have driven enterprise specialization in both crops and livestock. Coupled with government commodity programs which encourage continued production and high yields of so-called "program crops," this specialization in crop production has resulted in intensified per-acre chemical use (Runge et al., 1990). In some production regimes, this has further led to environmental problems, particularly those associated with ground water pollution.

Incentives for continued specialization and for intensive use of chemical pesticides and fertilizers will remain high for a number of crops. As a result, the environmental benefits could be high for biotechnology products which can substitute for these inputs in crop production. Some of these products will be induced by pressure from the larger society in the form of banned agricultural chemicals.

Livestock

As for crops, the U.S. livestock sector breaks down into two components. One is increasingly space-concentrated, higher technology, and intensively managed. This includes specialized cattle feedlots, broiler and swine production under confinement, and some large, highly specialized, dry-lot dairy operations. The latter dominate dairying in the Southwest and Florida but are also interspersed throughout several other production regions.

A second component of the livestock sector is the range livestock sector (which includes a large number of beef cow-calf units and a much smaller sheep segment), along with a variety of small, lower technology livestock enterprises, many of which are operated by part-time farmers. Many of these are concentrated in the Southeast and Appalachian regions of the nation.

This above-described categorization of the livestock sector into two components is an oversimplified but important one. A variety of biotechnology applications are, if approved for use, expected to have rather substantial adoption rates within the higher technology component of the livestock sector but not within the lower technology, spatially dispersed segment. In addition to the difficulty of administering some biotechnology products and processes except under conditions of space confinement and intensive management, the managers of small farms are less likely to incur the substantial learning costs required to effectively employ advanced biotechnologies.

The Case of BST

The use of BST (bovine somatotropin) in milk production provides an instructive example. Major increases in milk production (10 to 15% or more) are potentially achievable using BST after the initial post-freshening stage of lactation (Fallert et al., 1987). But effective utilization of the potential, including reduced production costs, will require rather intensive monitoring and management of feeding rates and rations, milk production, and animal health and breeding programs, along with high-level herd and system management.

And, as suggested earlier, operators of smaller dairy herds with lower per cow production levels are least likely to even undertake the adoption of BST. For those operators of larger, higher producing dairy enterprises who do adopt BST, their action will widen even further the per cow production-profit gap with lower technology producers. This is likely to hasten the exit of many of the latter group from milk production.

Projected adoption rates for BST technology, if and when it is approved for use, differ widely (Kalter et al., 1985; Marion et al., 1988). Moreover, the economic consequences to producers from increased milk supplies are shaped by the specific dairy programs which are in effect at the time of biotechnology adoption (e.g., price support, classified pricing, and surplus removal programs). Moreover, since higher technology dairy producers have some degree of regional concentration, these regions may gain further competitive advantage should BST be approved and adopted widely.

BST has become a salient public issue in some quarters. Opposition to the use of the technology comes from several perspectives (Molnar et al. 1990). Included are objections on the basis that milk supplies are already in surplus, concerns about the application of biotechnologies to living animals, perceived adverse impacts on small producers, and others.

PST may lead to a major restructuring in the swine sector, particularly in the direction of increased integration between producers and meat packers.

Other Examples

In the case of swine production, a high percentage of hogs are already produced on farms using specialized confinement systems. However, if PST (porcine somatotropin) is approved for commercial use, sizable improvements in feed efficiency and desirable changes in product quality (decrease in carcass fat and increase in lean meat content) are expected to result (Hayenga, 1988). Such product quality improvement could bring both incentives and pressures to differentiate these improved products in the retail marketplace. This differentiation could, in turn, occur via market grades or proprietary (branded) products or both. In any event, producers of non-PST-produced hogs might suffer a major setback in the marketability of their product.

PST may lead to a major restructuring in the production sector for swine. Possibilities include increased integration between producers and meat packers, possibly including PST suppliers. If PST is approved for use, the economic pressures will probably be strong for most swine producers to adopt this new technology or exit the industry.

In the case of both dairy and pork products, there will likely remain a significant consumer population that will prefer to purchase products produced without the use of growth hormones. The size of this market-demand segment is hard to estimate but will undoubtedly support at least limited production without hormone use.

A number of other biotechnology products, such as improved disease vaccines, can probably be utilized effectively by most livestock producers and will not have significant new management requirements for adoption and use. The extent to which such innovations are commercialized and adopted will depend on their profitability and effectiveness compared to that of other available technologies.

One consequence of increased specialization in livestock production has been a shift to larger and fewer enterprises on individual farms. Improved animal-health-control technologies have been a major contributor to successful production under confinement. Animal waste is concentrated by such systems, creating a major environmental issue in some locales. Thus, in specialized animal production, there is a continuing need for technologies that enhance waste disposal and utilization.

In summary, expected biotechnology innovations will probably support and encourage a continuation of the secular decline in the relative importance of small, spatially dispersed producers as suppliers of most livestock and livestock products. And, where product quality is influenced strongly by biotechnologies, as may well be the case for pork, increased incentives for production-marketing linkages via contracting and other forms of vertical coordination can be expected.

Forestry

Given the extended reproductive cycle of a tree, the major benefit of biotechnology lies in the expanded ability to select, propagate, and modify the performance characteristics of new varieties. A common characteristic of most trees is the long time period between planting and harvesting or other beneficial use. This time span varies substantially, however, between species and among end uses.

Most indigenous wild-type forests will probably remain genetically unaltered in perpetuity because of economic and forest management constraints. "Wild-type forests will serve as gene pools for species preservation, gene reserves for improvement programs, and major sources of forest products and recreation" (Haissig et al., 1987). But they will probably be affected very little by the emerging biotechnologies.

Emerging biotechnologies will favor urban/ornamental forestry and commercial forestry for pulp and lumber production over expansive, wild-type forests and small farm woodlots.

As in the case of agriculture, commercial forestry has several somewhat diverse components. These range widely from nurseries for urban and ornamental tree production, to highly managed forests for pulp and lumber production, to numerous small farm woodlots, to expansive wild-type forests. The relative importance of this last segment in the production of forest crops is already diminishing and will likely continue to do so in the future.

Many small farm woodlots serve one or more multipurpose roles, including windbreaks, source of firewood supplies, wildlife habitat, recreation, and source of commercial forest products. Among woodlots, it is only in selective cases of commercial production of forest products that biotechnology applications are likely to have significant adoption. Thus, urban and ornamental forestry and the commercial production of wood and paper products will gain the most from the emerging biotechnologies.

Among the list of application possibilities for biotechnologies in forestry are those of producing wood products of superior quality; faster and higher quality tree growth; improved tolerance to diseases, insects, and other environmental stress (drought, poor soil drainage, frost, etc.); and superior ornamental attributes (shape, size, color, etc.). Also of key importance to tree growers is the enhanced capability via biotechnology applications for researchers to shorten tree breeding time, the time between genetic test establishment and evaluation and between selection of improved individual trees and their reproduction in plantable quantities. This will permit commercial tree growers to capitalize on genetic gains within a shorter span of time.

One of the corporate giants in the forest-products industry now produces all of its tree seed in orchards managed with traditional and advanced (but non-molecular) pollination techniques. It is from this advanced stage of quality control and management that new biotechnology innovations will be applied to this component of the forestry sector. Of central importance is the expected widespread use of tissue culture procedures to supply genetically improved genotypes and tissue culture micropropagation to supply the means of producing clones for direct commercial application and breeding programs.

Structural impacts of biotechnologies are not expected to be of major consequence for the forestry sector in the near term. But, over time, the use of biotechnologies is likely to have the overall effects of (1) improving the productivity of resources used in commercial forestry, (2) improving the overall quality of forest products, and (3) reducing the importance of small-scale woodlot producers and wild-type forestry as commercial suppliers of forestry products.

Small-scale producers will depend on a broad-based market availability of biotechnology products and processes. They also will require low-cost guidance pertaining to the use and management of these technologies. The information availability requirement is particularly likely to require effective public education and technical assistance programs.

Aquaculture

Water-based husbandry of aquatic species encompasses a diverse set of production environments and probably represents one of the most globally integrated food-production sectors. Because the world catch of fish has reached biological limits in the neighborhood of seven million metric tons, aquaculture remains the only alternative for increasing global fish supplies.

A central challenge to each of the multitude of specific industries encompassed by aquaculture is to understand and control the organism's reproductive process. Many species of fish, shrimp, clams, oysters, mollusks, eel, and squid are grown in warm, cold, fresh, salty, and brackish waters. Some species have been cultured for over 2,000 years in China, while the means for artificially spawning shrimp have been perfected

Biotechnology represents a new means of accelerating technical progress in aquaculture.

only in the past decade. Such breakthroughs have proven to be particularly timely given the seemingly insatiable world demand for the commodity.

Clearly, biotechnology represents a new means for accelerating technical progress in aquaculture by circumventing bottlenecks in seedstock production, ensuring uniformity and quality in the biological base of these industries. Many animals of commercial importance to aquaculture lend themselves well to genetic manipulation because of their high fecundities and external fertilization. Their reproductive cycles are readily controllable by manipulations of water temperature and photoperiod or by administration of reproductive hormones. In many cases, their embryos can be artificially incubated in a controlled environment. These reproductive features facilitate the making of various crosses for purposes of selective breeding and hybridization. Moreover, the generation time in some species is quite short, allowing rapid improvement of strains through selective breeding or other genetic manipulations.

The main potential benefits for aquaculture via biotechnology lie in facilitating the identification and propagation of strains superior in terms of disease resistance, reproductive vigor, feed efficiency, size, and desirable product-quality characteristics. Trout, salmon, and catfish are examples of species cultured in high volume where biotechnology research is in progress. Biotechnology offers the clear prospect of accelerating the search for desirable production traits, handling characteristics, and marketing attributes, and of hastening the transition of seafood production from a hunting-gathering activity to a farm-based industry.

Practical use of transgenic fish in aquaculture awaits the development of cost-effective methods for gene transfer and selective breeding. Moreover, considerable uncertainty surrounds the question of environmental effects associated with the accidental escape of transgenic fish from aquaculture operations. Application of recently developed sterilization methods, including chromosome and hormone manipulation, has been proposed as a means of reducing environmental risk. Although biotechnology innovations will eventually accelerate species development, most of these applications are still some years away and will have little near-term impact on organization and operation of the aquaculture sector.

In the longer term, however, successful participants in the commercial aquaculture sector, particularly in fish production, will need to utilize a variety of biotechnology procedures. As in the case of agriculture and forestry, availability of appropriate low-investment technology and an effective public information system will be important if small-scale producers are to compete effectively. And pressures will increase for vertical coordination and integration in a food sector in which successful producers must already develop close contractual linkages with marketing outlets.

Consequences of Biotechnologies for the Input Supply Sector

To date, the largest impacts of the emerging biotechnologies have not been on agricultural producers but on firms manufacturing production inputs.

Although no biotechnology-enhanced seed is yet available for large-acreage volume crops, the expectation of early commercialization of such seed has already altered the structure of the seed and chemical industries (Molnar and Kinnucan, 1989). Restructuring of the hybrid seed-corn industry is the early and leading example.

Since the 1930s, the rapid adoption of hybrid-seed corn has fueled the development of that seed industry to an annual billion-dollar business. Because the offspring of hybrid seed are not reliable, farmers were guaranteed to be repeat customers, and profits flowed steadily to the seed industry. The sustained past profitability of seed companies, together with expected future gains through biotechnology, has led to their acquisition by multinational petrochemical and pharmaceutical corporations. Of the major seed companies, only Pioneer Hi-Bred International remains an independent firm (Kloppenburg, 1988).

Increased concentration (fewer firms) in the seed and other input industries also raises the potential for monopoly power. Potential monopoly power is a concern for two reasons: the potential for exploiting farmers and other producers in their purchases of improved inputs and ecological vulnerability stemming from overdependence on a narrow set of genetic material (Doyle, 1985). In support of their efforts to privatize the benefits from new improved hybrids, the commercial seed industry is exerting pressure on the public research sector to focus on "upstream" grandparent line development and to leave the "downstream" hybrid variety work to the private sector.

Proprietary production processes and prescribed production regimes associated with contracting and direct vertical integration could result in a captive market for some biotechnology products. Moreover, some of the more technically sophisticated biotechnology products will likely require new facilities and/or new skills for employees of supply firms operating at retail. Some firms will choose not to invest in the additional cost of obtaining such resources and, in the process, will reduce or lose their capability to market some biotechnology innovations.

Economies of scale in process technologies provide additional potential for concentration in the input supply sector. For example, the output of a seven-million-dose-per-day BST plant is sufficient to treat about two-thirds of the nation's dairy herd. To the extent that biotechnology manufacturing requires large minimum plant sizes to be efficient, there will be economic pressures to concentrate industry structure to, at most, a small number of firms. Moreover, there may, in the cases of some production inputs, be increased incentives for manufacturing firms to integrate the manufacturing and retailing of these inputs.

Private firms already involved in pharmaceutical research can easily move into animal agricultural biotechnologies. Dairy, beef, swine, and poultry will be the leading markets for biotechnology products in the near future. Although the dominant scientific techniques used in animal biotechnologies do not involve release of genetically engineered material into the environment, health- and food-safety issues remain as potential hurdles for both animal and plant products.

The above-mentioned possibilities for increased importance of proprietary products and processes in the input supply sector and the economic incentives for further industry concentration suggest three requirements in support of small-scale producers of food and forestry products. First, there will be a need for directing public research to low-investment technologies which they can use effectively in their production systems. Second, they will need economical market access to these technologies. Third, they will need low-cost information sources on use and management of available technologies.

Impacts of Biotechnologies on Rural Communities

As the U.S. farm population has declined, many smaller farm-supply and marketing businesses have also disappeared, along with numerous consumer retail stores and banks. As a consequence, many agriculturally dependent rural communities have lost population, while some larger trade-center communities have grown both in population and in business volume. Moreover, these growing communities often are preferred locations for consolidated public schools, medical facilities, and other public services. The communities left behind are suffering the consequences of changing time/distance relationships associated with better roads and communications, and some places are particularly vulnerable to a changing agricultural structure.

Increased demand by producers for one-stop shopping centers for farm supplies and technical services (including those involving biotechnologies) may reduce even further the viability of agribusiness enterprises in smaller rural communities. As a result, some rural places will need to substitute additional nonfarm economic activities if they are to replace the receding sustenance base associated with agriculture. But that would generally be true with or without the emerging biotechnologies.

Precise connections between the emerging biotechnologies and the structure and viability of rural communities are difficult to identify and measure. However, the impact of agricultural change on rural communities is, to a high degree, proportional to the level of local dependence on agriculture. Fewer than 40 congressional districts nationwide now have more than 20 percent of their population living on farms. These counties are concentrated in the Great Plains region of the country, which has a low density of population and limited nonfarm economic activities. Insofar as biotechnology is a source of structural change in agriculture, the incidence of change will probably be closely related to innovations affecting the production of specific commodities. Thus, in the near term, any community effects will likely be most significant in selected regions of concentrated livestock production. The ability of rural communities in these regions to absorb adverse changes in agricultural employment will be closely related to the availability of off-farm employment (Office of Technology Assessment, 1985).

But biotechnology is only one of an array of forces shaping local economies and labor markets. Rural communities in the Midwest have been most severely affected by the recent non-biotechnology-induced financial crisis in agriculture and remain vulnerable to future adversities, whatever their source.

Biotechnology innovations can be expected to draw new attention to agricultural production practices. Several events, including the initial field testing of ice minus bacteria in California and community-based resistance to BST in the Midwest, suggest the vulnerability of some biotechnology innovations to adverse community opinion.

Growing numbers of community residents have little or no personal contact with farming, save as passive observers of changes in the environment. The environmental impacts of production practices can, however, become a community issue when externalities such as water quality, chemical residues, worker safety, etc., become sources of concern. Biotechnologies have already raised local sensitivities about the implications of novel substances employed in producing crops or animals. Perceptions of risk to health, safety, and/or environmental diversity associated with transgenic organisms may become a further source of community conflict and controversy.

At the community level and beyond, output-increasing biotechnology innovations may reinforce the trend to fewer, larger firms producing food, fiber, and forestry products. There are no clear guidelines as to the optimal organizational structure for these sectors, though there is widespread public support to preserve the family farm (Comstock, 1987; Friedberger, 1988). Since much of the public concern about effects of biotechnologies centers on employment opportunities and the general economic viability of rural communities (the latter including the economic viability of smaller production firms), it is primarily to these topics that public information and debate on the structural implications of biotechnologies (and possible remedial policies) need to be targeted.

Finally, from a rural community standpoint, is the question about the expected impacts of biotechnology innovations on the "quality of life." As a value-laden question, this issue is difficult to answer because of a lack of agreement as to what constitutes quality of life. For some, income levels are a major determinant of quality of living. For others, environmental quality and/or the incidence of family farming may be crucial quality determinants. Changes that widen inequality

. . . biotechnology is only one of an array of forces shaping local economies and labor markets.

or cause rapid change in the livelihoods of some community residents also undermine the quality of life in a locale. "Sustainability" of agriculture is a recent addition to the list, and there are others.

Probably of key importance in assuring adequate consideration of the quality of life relative to the impacts of biotechnology innovations are:
(1) open public discussion regarding biotechnology research priorities;
(2) enlightened policies and procedures regarding the approval, patenting, and regulation of resulting innovations; and
(3) insistence on adequate high-quality and timely information about biotechnologies for both public and private decision makers. If these outcomes can be achieved, most quality of life objectives may be achieved also.

Conclusions

Biotechnologies are already affecting some components of the agricultural supply industry in a major way, and their impact on the production sectors for food and forest products will grow over time. Biotechnology is probably not likely to be the bane or boon its strong detractors and promoters make it out to be. It can be asserted, however, that biotechnology will be a significant factor driving the evolution of future agricultural production systems. Specific developments will have substantial transforming effects on the production and input supply sectors. And they can result in needed productivity gains and environmental benefits. Other biotechnologies will have only subtle or marginal impacts on production systems but can benefit producers, processors, and/or consumers.

Biotechnologies are not inherently detrimental to a family-farm structure of agriculture or to smaller production firms in forestry and aquaculture. But such family-scale production systems will require the support of targeted research and development, of market access, and of information services designed for their needs. This will often require the development and availability of technologies which can be used effectively in conjunction with low-investment production systems.

Biotechnology, however, is only one of an array of shifting technological frontiers altering production practices. Advances in computing and information processing increase the quantity, timeliness, and accuracy of information available to the producer, input supplier, and processor. Improvements in irrigation, machinery, and material handling, not to mention the use of robots, have many yet-to-be-realized consequences for the food, fiber, and forestry systems. Corollary developments in agricultural markets use new financial instruments to shift risk and increase the pricing and operational efficiency of the food marketing system.

In addition, changing markets, national and international policies, and international competition will affect the future organization and welfare of producers and input suppliers. Thus, biotechnology is only one of many sources of change for agriculture, forestry, and aquaculture.

Safety and environmental issues aside, many populist explanations for the recent difficulties experienced by farmers and agriculture generally cite much of the source of the problem as government involvement in agriculture and actions of large corporations thought to exploit the farmer. Though many of the innovations are emanating from small private firms and university laboratories, much of the input manufacturing and distribution apparatus is in the hands of a few large multinational corporations that have the financial wherewithal to develop and market high-technology products.

At the rural community level, as well, biotechnology is only one of many factors affecting structural change and public opinion. Effects of technological change will continue to be greatest in those communities where agriculture is an important source of income and employment. Those communities in close proximity to agriculture will also be the most concerned about environmental and safety issues related to biotechnology. Effective mechanisms for credible public information on these issues will be important. So will the facility for community input into discussions and decisions about technology development, implementation, and regulation.

References and Additional Readings

Cochrane, W.W. 1979. The development of American agriculture. University of Minnesota Press, Minneapolis, MN.

Comstock, G. 1987. Is there a moral obligation to save the family farm? Iowa State University Press, Ames, IA.

Doyle, J. 1985. Altered harvest. Viking Penguin, New York, NY.

Fallert, R., T. McGuckin, C. Betts, and G. Bruner. 1987. bST and the dairy industry: A national, regional, and farm-level analysis. U.S. Dept. of Agric., Economic Research Service, Agricultural Economic Report No. 579.

Friedberger, M. Farm families and change in 20th century America. The University Press of Kentucky, Lexington, KY.

Genetic Engineering News. 1989. Agbio industry positions itself to bring products to the marketplace in the 1990s. Vol. 9, No. 10, November/December.

Haissig, Bruce E., Neil D. Nelson, and George H. Kidd. 1987. Trends in the use of tissue culture in forest improvement. Biotechnology, Vol. 5, January.

Hassebrook, C and G. Hegyes. 1989. Choices for the heartland: Alternative directions in biotechnology and implications for family farming, rural communities. Center for Rural Affairs, Walthill, NE.

Hayenga, Marvin L. 1988. Biotechnology in the food and agricultural sector: Issues and implementations for the 1990s. UC AIC Issues Paper No. 88-5, Agricultural Issues Center, University of California, Davis.

Hess, C.E. 1987. Agricultural biotechnology. National Academy Press, Washington, DC.

Kalter, R.J. et al. 1985. Biotechnology and the dairy industry: Production costs, commercial potential and the economic impact of the bovine growth hormone. A.E. Research 85-20, Dept. of Agricultural Economics, Cornell University.

Kloppenburg, J. 1988. First the seed: The political economy of plant biotechnology. Cambridge University Press, New York, NY.

Marion, B.W., R.L. Wills, and L.J. Butler. 1988. The social and economic impact of biotechnology on Wisconsin agriculture. Report to the State Legislature. College of Agricultural and Life Sciences, University of Wisconsin, Madison.

Molnar, J.J. and H. Kinnucan, eds. 1989. Biotechnology and the new agricultural revolution. Westview Press and the American Association for the Advancement of Science, Boulder, CO.

Molnar, J.J., K.A. Cummins, and P. Nowak. 1990. Bovine somatotropin: Biotechnology product and social issue in the U.S. dairy industry." Journal of Dairy Science. Vol.73.

Office of Technology Assessment. 1985. Technology, public policy, and the changing structure of American agriculture. OTA-F-272. Office of Technology Assessment, U.S. Congress, Washington, DC.

Reilly, J. 1987. Cost-reducing and output-increasing technologies. USDA-Economic Research Service Technical Bulletin 1740, Washington, DC.

Rodgers, E.M. 1982. Diffusion of innovations. The Free Press, New York, NY.

Runge, C.F., R.D. Munson, E. Lotterman, and J. Creason. 1990. Agricultural competitiveness, farm fertilizer and chemical use, and environmental quality: A descriptive analysis. Center for International Food and Agricultural Policy, University of Minnesota, St. Paul, MN.

Teigen, L.D. 1986. Technological diffusion: Effects on productivity, structure, firms, and markets. p. 361-390. In Agricultural change: Consequences for southern farms and rural communities, Westview Press, Boulder, CO.

Tiedje, J.M., R.K. Colwell, Y. Grossman, R. Hodson, R. Lenski, R. Mack, and P. Regal. 1989. The planned introduction of genetically-engineered organisms: Ecological considerations and recommendations. Ecology 70, April, 2:298-315.

CHAPTER THREE

Biotechnology has revolutionized biology and is destined to have an even greater impact than the industrial revolution on the agriculture and food systems in this country. Farmers and food processors have had centuries of experience in applying "old" biotechnology techniques. Domestication of plants and animals, selection of superior food-producing varieties and breeds, and fermentation technology, including the use of microorganisms for the production of a vast array of fermented foods, are examples of traditional biotechnology practices.

The "new" biotechnology, which has emerged since the early 1970s, expands upon our ability to improve plants, animals, and microorganisms. It could dramatically improve the productivity and efficiency of food production and processing, and expand and extend food and nonfood uses of raw agricultural commodities.

Creative application of the tools of biotechnology will be required to assure a safe, wholesome, nutritious, and affordable food supply for future generations. The challenge will be to develop policies which balance society's need for a safe and abundant food supply with the possible socioeconomic effects that could occur should biotechnology be broadly applied to all aspects of the food chain.

3 Impact of Biotechnology on Food and Nonfood Uses of Agricultural Products

Susan K. Harlander, University of Minnesota
James N. BeMiller, Purdue University
Larry Steenson, Purdue University

Introduction

The "Old" vs. the "New" Biology

Biotechnology is not new to the food and agricultural sectors. We have been consuming foods produced through biotechnology for centuries. Cheese, sausage, pickles, bread, beer, wine, and a wide variety of other fermented food products are made with the help of bacteria and yeast. As our understanding of the metabolism of microorganisms has expanded, we have learned how to use these organisms for the production of many food ingredients, including sweeteners (aspartame and xylitol), stabilizers (xanthan gum), flavor enhancers (monosodium glutamate), and nutritive additives (vitamins and amino acids).

Domestication of crops and animals, together with classical breeding and selection of superior varieties and breeds, are well-accepted practices for improving the food production system. These traditional forms of biotechnology have been practiced and refined for thousands of years.

We have directed the genetic improvement of food-producing plants, animals, and microorganisms for all of this century and have done it very well. Using the tools of molecular biology which have been developed within the last 20 years will allow single traits to be modified in a much shorter time frame, and with more predictable and controllable outcomes, than would be possible with traditional selection and breeding methods.

The new biotechnology will affect the entire food chain, from the planted seed to the consumer's table. Biotechnology has the potential to reduce the need for agricultural chemicals; improve the productivity, efficiency, and profitability of food production; open new markets for improved or unique food and nonfood products; and improve the quality, safety, cost, and convenience of consumer food products.

Chapter Concerns

This chapter explores the potential implications of biotechnology on food and nonfood uses of biotechnology. The first part of the chapter focuses on how biotechnology can be used to improve the utilization of raw agricultural commodities as food for human consumption. The primary emphasis is on biotechnological processes that can be applied after the commodity leaves the farm gate.

The second part of the chapter focuses on the potential impact of biotechnology for expanding nonfood industrial uses for raw agricultural products. It is important to note that many of the applications discussed here are theoretical, because the food-processing and commodity chemical industries have been slow to apply biotechnology.

The socioeconomic context within which the agricultural and food processing industries function has often been incompatible with the introduction of new, innovative technologies. Great change requires adjustment and can cause dislocation. Thus, the third part of the chapter focuses on some of the socioeconomic issues surrounding the adoption of new biotechnological developments and what effects this might have on the structure of the food and agricultural system in this country.

Impact of Biotechnology on Food Uses of Agricultural Products

Impact of Biotechnology on Production Agriculture

The first impacts of biotechnology on the food system will be in the production agriculture sector. There are numerous examples of disease-resistant, herbicide-tolerant, and insect- or virus-resistant plant varieties developed via selection in tissue culture or by genetic engineering techniques. These same methods have been used to develop salt-tolerant, temperature-tolerant, and drought-resistant crops. A microbial gene from *Bacillus thuringiensis* has been engineered into several plant species, including tomato, tobacco, and cotton. The first commercially available engineered plants will involve single gene modifications in well-understood metabolic pathways.

As illustrated in Table 1, many engineered plants are currently being tested in field trials. The ultimate goals are to reduce dependence on chemical pesticides and herbicides and to increase profits for farmers. It may be possible to increase the productivity and efficiency of production agriculture.

Custom-Designed Raw Ingredients

The food processing industry transforms raw agricultural commodities into the shelf-stable, convenient, and palatable processed foods found on supermarket shelves. Historically, food processors have viewed raw agricultural crops as generic commodities. Although some vegetable, fruit, and grain processors are involved in breeding programs aimed at developing varieties that meet their specific needs, most processors have little choice but to process what is available, modifying the manufacturing parameters in order to compensate for inconsistencies in raw material quality.

Application of the tools of biotechnology offers the potential to custom-design raw commodities with improved nutritional, functional, or processing characteristics which provide added value for the processor. This allows the food processor to design and engineer proprietary products to fill a specific market niche. Examples of successful application of biotechnology for construction of crop varieties with improved properties are listed in Table 2.

In most cases, the properties to be modified are controlled by many genes and complex metabolic pathways which are not well understood. Much more fundamental information on plant biology must be available before custom-designed agricultural crops can be constructed. However, one can imagine what the future might bring.

As we understand more about the role of diet in health and disease, one can imagine using the tools of biotechnology to enhance the level of specific nutrients or components, such as soluble or insoluble fiber, or specific vitamins and minerals associated with healthier foods. The bioavailability, or digestibility, of nutrients could be enhanced, and natural toxicants, or antinutrients, in foods could be eliminated. Fruits and vegetables with improved flavor, texture, aroma, and shelf life

Table 1. *Field Trials of Genetically Engineered Plants.*

			Field trials	
Crop	**Nature of change**	**Advantage**	**Approval**	**Location**
Tomato	Tolerance to glyphosate	More environmentally safe herbicide	12/87	California
Tomato	Resistance to alfalfa mosaic virus	Disease resistance	5/88	Wisconsin
Tomato	Expresses delta-endotoxin gene of *B. thuringiensis*	Resistance to certain insects	2/89	Florida
Potato	Resistance to leaf roll virus, virus x, and virus y	Disease resistance	4/89	Illinois
Cotton	Tolerance to glyphosate	More environmentally safe herbicide	5/89	Alabama
Soybean	Tolerance to glyphosate	More environmentally safe herbicide	5/89	Tennessee

Source: Morris and Przybyla, 1989

Table 2. *Custom-Designed Crops for the Food Processing Industry.*

Crop	Nature of Change	Impact on Processing
Tomato	Increased solids content	Decreases cost for transport, less water to remove during processing
Carrots	Sweeter, crunchier	Longer shelf life
Corn	Increased level of specific amino acids	Enhanced nutritional quality
Rapeseed	Decreased level of saturated fatty acids	Enhanced nutritional quality
Tomato	Decreased levels of endopolygalacturonase	Longer shelf life, vine-ripened flavor, decreased need for refrigeration during transport

Source: Morris and Przybyla, 1989

would encourage the inclusion of more fresh produce in the diet.

As we understand more about how the structure of a raw food ingredient affects functional properties during processing, one can imagine using tools of biotechnology to engineer crops for specific processing applications. What about a starch that doesn't stale for use in breadmaking or a variety of corn that produces nonsodium salty peptides for making corn chips or other salty snacks? Rather than viewing corn as a generic commodity, food processors could select the custom-designed variety most suitable for their particular processing application.

Plant Cell Tissue Culture

Plants serve as the source of many ingredients used in processed foods. These include flavors, pigments, essential oils, stabilizers, and enzymes, which are normally extracted from intact plants. Plant cell tissue culture offers an alternative to whole plants as a biological source of useful compounds. Tissue can be excised from the root, stem, leaf, or fruit of plants, and the undifferentiated cells can be propagated in solid or liquid media containing all of the essential nutrients required for growth.

Plant tissue culture for production of natural food ingredients offers several distinct advantages over extraction of these components from plants. Seasonal variations, unfavorable weather conditions, and epidemic diseases are not problems when plant tissue is grown under well-defined and controllable laboratory conditions. Because many of the plant-derived ingredients of interest must be purchased from foreign countries, food processors must sometimes deal with an erratic supply due to natural or political calamities in these countries. In addition, they have little control over how plants are grown, harvested, transported, or stored.

Plant cell culture allows the processor to control the quality, availability, and processing consistency of the ingredients. The successful production of the purple dye, shikonin, by plant cell suspension cultures illustrates the potential for biosynthetic production of plant products of interest to the food industry. Examples of other high-value food ingredients which could be produced by plant cell suspension cultures are listed in Figure 1.

Plant tissue culture technology provides many other applications of interest to the agriculture and food sectors. Tissue culture technology is used for large-scale production of superior genetic varieties for commercial use. Propagation via tissue culture may allow for the rapid development of improved varieties and replace time-consuming conventional breeding procedures. In addition, this technology is used for propagation of elite plants (and hybrids) that are difficult to produce using conventional seed technology, for the production of virus-free nursery stock, and for the propagation of certain high-value perennials, fruit trees, estate crops, and forest trees.

In contrast to clonal propagation described above, which faithfully produces genetically identical plants, regeneration of plants from callus, leaf tissue explants, or plant protoplasts results in recovery of somaclonal variants. Somaclonal variation can be used to select plants that exhibit changes in fruit color, plant architecture, or characteristics that improve the

Figure 1. *Examples of Potential Food Products from Plant Tissue Culture.*

Food Colors
anthocyanins
betacyanins
saffron

Vegetable Flavors
tomato
celery
capsicum
asparagus

Fruit Flavors
raspberry
strawberry
cherry
grape
pineapple
apple
peach
apricot
banana

Nonfruit/ Nonvegetable Flavors
vanilla
cocoa

Oils
mint
garlic
lemon
jasmine
rose
onion
vetiver
patchouly

Spices
cinnamon
cardamon
rosemary
sage
tumeric

Nonnutritive Sweeteners
stevioside
thaumatin
monellin

ease of mechanical harvesting. Variants can be challenged with disease agents in cell culture in order to select disease-resistant varieties. The ability to culture haploid plants from pollen cells which contain half the genetic information found in normal plant cells allows one to identify the best characteristics of each parent for the generation of true-breeding recombinant lines.

The discovery of protoplast, or plant cell fusion, permits the combination of plant cells of different genetic makeup and leads to the production of new hybrid plants with unique characteristics. New breeding lines have been produced between cultivated crops and closely related disease-resistant wild species. The hybrids produced by protoplast fusion contain a mixture of genetic information from each parent, and, although the process is time-consuming, it opens the door to the transfer of new traits that are not accessible using conventional breeding approaches.

Genetic Improvement of Food Fermentation Microorganisms

Microorganisms have been used for centuries for the production of fermented foods, such as cheese, sausage, sauerkraut, wine, and bread. Genetic engineering provides an alternative to classical mutation and selection for improving microbial starter cultures with improved metabolic properties.

Genetically engineered starter cultures are not yet commercially available; however, some examples of how genetic engineering could be used to improve organisms for various fermentation processes are provided in Table 3. Availability of such strains would have an impact on several aspects of fermentation, including production economics, shelf life, safety, nutritional content, consumer acceptance, and waste management.

Microbial Ingredients Used in Food Processing

Microorganisms can also be used for the production of ingredients for processed foods. Consumer concern regarding chemical additives in foods and consumer demand for "natural" products have resulted in increased interest in microbial metabolites which can be used as natural ingredients in foods.

A number of food ingredients are already being produced by microorganisms, including sweeteners (aspartame), flavors (buttery aroma, malt flavor) and flavor enhancers (MSG), vitamins (B12), amino acids (leucine, phenylalanine), acidulants (acetic and citric acid), thickening agents (xanthan gum, dextrans), and antimicrobial agents (nisin). In many cases, agricultural commodities or waste products generated by the food industry can be used as feedstock for microbial production of food ingredients.

Microbially derived enzymes are used extensively by the food-processing industry, which accounted for over 50 percent of world enzyme sales in 1987. Enzymes perform many valuable functions in food systems. They help control texture, appearance, and nutritive value, as well as the generation of desirable flavors and aromas. Most of the enzymes are hydrolases that degrade large molecules, such as proteins, carbohydrate polymers, or lipids, to their component parts.

Most enzymes are not optimally functional under conditions used during food processing (high temperatures and acidic pHs). Protein engineering provides the necessary tools for modifying the substrate specificity, pH and temperature stability, and resistance to protease digestion. Improved enzymes will expand the uses for enzymes in food processing and increase the kinds of raw materials which can be utilized as food for animals and humans.

Animal Biotechnology

Growth hormone genes from several animal species have been isolated, characterized, and cloned into microorganisms for the efficient production of large quantities of pure protein. When injected into dairy cows, bovine somatotropin (BST) has been demonstrated to increase milk production and feed efficiency and to accelerate the growth rate of animals.

BST is a peptide hormone which is normally present in milk in small quantities. The level of BST in milk from treated animals is within the range normally found in milk from untreated animals. Because it is a protein, BST is digested in humans and, as a species-specific hormone, is nonfunctional in humans, even if injected.

BST is a management tool for farmers and could be used to improve the productivity and efficiency of milk production. BST is currently under review by the U.S. Food and Drug Administration and will likely be one of the first products of biotechnology to affect the food chain.

PST, the porcine equivalent of BST, stimulates growth rate and feed efficiency of pigs and also has the ability to dramatically alter carcass composition, significantly decreasing the amount of fat and increasing the amount of protein. Meat from PST-treated animals is leaner than meat from untreated animals and might be more desirable to health-conscious consumers.

Table 3. *Genetic Improvement of Microorganisms.*

Type of fermentation	Nature of improvement	Benefit
Dairy	Bacteriophage (virus) resistance	Eliminate economic losses due to destruction by virus infection
	Accelerated ripening of cheese	Decreased storage costs
	Higher levels of the enzyme beta-galactosidase	More digestible for lactose-intolerant individuals
Meat	Bacteriocin production (natural preservative)	Inhibition of pathogenic or spoilage organisms
	Addition of cholesterol-reducing enzymes	Reduction of an undesirable dietary component
	Addition of fat-modifying enzymes	Alteration of the saturated to unsaturated fat ratio
Beer	Alpha-amylase production	Production of "lite" or low-calorie beer

Growth hormones from other animals like fish and chickens are also currently under investigation. These applications will have a profound impact on the efficiency and profitability of animal agriculture.

In the future, other properties of animals could theoretically be altered using genetic engineering techniques. As we understand more about the mechanisms which regulate disease resistance, reproductive functions, body composition, and partitioning of components like fat, cholesterol, and protein, it may be possible to modify these properties to improve production efficiency and nutritional value. This could increase the market value of these products for farmers.

Aquaculture, or fish farming, has become increasingly popular as an alternative to commercial fishing for the propagation of certain species of fish. Dietary and environmental conditions can be carefully controlled, eliminating concern about consumption of fish from contaminated lakes, streams, and oceans. In addition, varieties of fish can be grown year-round at any location. The availability of fresh fish year-round could dramatically influence consumer eating habits, creating an increased market for fish. This could affect the demand for chicken and red meat.

As we unravel the mysteries of how diet affects health and chronic disease, biotechnology could be used to custom-design the food supply to meet the nutritional needs of specific target populations. Ultimately, it will be possible to determine each person's nutritional requirements and design an individualized diet to meet his or her needs.

Impact of Biotechnology on Nonfood Uses of Agricultural Products

For centuries, replenishable products of the farm and forest have been used for other than human food or animal feed. Examples include cotton for fiber and absorbants; wood cellulose for a wide variety of products, including paper, cellophane, rayons, plastics, and substances that improve the action of detergents; corn and potato starch, which can be used as is, as adhesives, and as binders for ceramics and can also be used to make insulating foams, to size and finish textile's, and to make products used to coat paper; and vegetable oils to make such things as lubricants, paints, and varnishes. Cellulose, starch, and vegetable oils are products of plants. However, none are proteins, the molecules that can be changed most easily as a result of genetic engineering.

At one time, zein, a corn protein which can be spun into a textile fiber, had a promising industrial future. However, development of this product was halted in midstream in the mid-1950s by the introduction of polyesters. Soy protein is used in the manufacture of certain papers.

However, except for trees for wood and pulp, cotton and flax for linen and linseed oil, and other oilseed crops such as castor bean, few crops are grown specifically for nonfood uses. Generally, when agricultural crops achieve nonfood uses, what is being used is a commodity grown for animal feed or human food for which an industrial use has been found. Otherwise, what is used industrially is a byproduct such as corn cobs and citrus peels.

There is as yet little being done to apply biotechnology in the development of new crops to supply industrial products or raw materials. However, biotechnology promises new roles for agriculture, which can become not only a source of food and fiber but a provider of energy, materials, and specific chemicals.

Crops That Produce Value-Added Products

Historically, new plant varieties have been developed by classical breeding programs that involved germ plasm collection and preservation, screening for desirable traits, the use of various agents to generate mutants, and the development of hybrids. This approach has been used successfully to produce plants that are adapted to different soil types and climates, produce higher yields, produce products of improved quality, are resistant to diseases and insects, and have improved traits such as even seed maturity and shatter resistance.

Classical selection and breeding techniques have also been used to produce the mutant strain of corn called "waxy maize" (or waxy corn), which produces superior starch products for many food applications. High-amylose corn can be used to produce another specialty starch for the food industry. These are two examples of specialty crops being developed or grown for a specific food use.

There is, however, growing interest in the use of plants or plant products for the production of nonfood products. An example is kenaf, an annual, nonwood fiber plant currently being investigated as a fiber source for the manufacture of newsprint.

The ability to alter the genetic structure of plants using biotechnology offers the possibility of engineering specific changes in the nature of plant components such as starch and oil. However, these components are made by complex biosynthetic pathways involving scores of enzymes, and synthesis itself is limited to specific structures within cells.

This presents scientists with a formidable task. In order to genetically engineer changes in plant components, scientists must first identify the enzymes of the biosynthetic pathways, their specific compartmentations, and the flow of precursors. Then, they must identify target enzymes for insertion, removal, or modification, along with controls (regulatory elements) that ensure that they are expressed in the right compartment, in the right tissue, and at the right time. Yet the unprecedented possibilities promised by the prospect of inducing specific changes in certain oils and starches make the task well worth completing.

Oils

The major oils used by both food and nonfood industries are soybean, corn, cottonseed, rapeseed (canola), sunflower seed, castor bean, linseed, palm, coconut, and tung oils. Potential sources of new oils, i.e., potential new oilseed crops, are jojoba, evening primrose, borage, blackcurrent, crambe, sal, mowrah, and mango. The value of a new commercial oil would lie in a special property that suits it better for a particular food or industrial use.

The principal components of any oil are fatty acids, not in the free form, but chemically combined. More than 1,000 different fatty acids are known, yet only 7 comprise more than 90 percent of the total found in nature. The rare ones have unusual structures and are potentially useful, but they generally come from nonagricultural sources. To use agricultural crops as their source would require improvements in yield per acre, processing, and applications.

Potential applications of biotechnology in this area include altering proportions of fatty acids in current oilseed crops; production of oil containing a specific fatty acid(s) by bacteria, yeasts, fungi, or microalgae in fermenters; and modification of oils by enzymes produced by microorganisms in bioreactors.

Fuel can be produced from vegetable oils and animal fats using current technology, but the price of petroleum-based fuels must be high enough to make it economical to produce fuel in this way.

Carbohydrates

Starch and cellulose are used extensively in nonfood applications. Both are polymers, i.e., very large molecules made up of the simple sugar glucose (dextrose) joined together in chains. For industrial uses, starch and cellulose molecules are often modified through chemical reaction. Chemical modification provides a broad range of products with applications in many industries, including the chemical, food, pharmaceutical, cosmetic, detergent, paper, and petroleum-producing industries.

In addition to the examples already given, modifications can be made in starch, ordinarily soluble in water, to make it water insoluble so that it can be incorporated into plastic trash bags to make them biodegradable. And cellulose, ordinarily insoluble in water, can be modified to make it water soluble and useful for enhanced oil recovery.

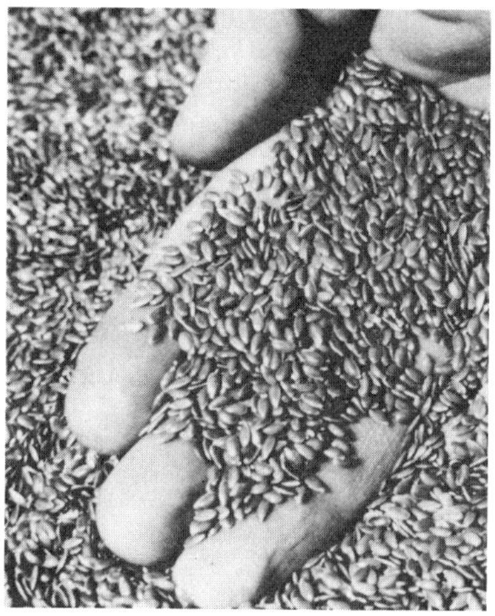

Biotechnology can help make specific changes in oilseed crops so that the oils are more useful for food and industrial purposes.

Sucrose (sugar) and, to a lesser extent, lactose (milk sugar), a food processing byproduct, are also plentiful and readily available carbohydrates. As yet, no major industrial use has been found for either.

Fermentation

Carbohydrates account for about 70 percent of the dry weight of all biomass (living organisms) and are a preferred source of energy for growing microorganisms (bacteria and yeast). Thus, glucose and corn syrup (made from corn starch), sugar, and other carbohydrates are used in fermentations to make a variety of useful products. Among these are ethanol (alcohol fuel); enzymes; water-soluble gums such as xanthan, which is used in tertiary oil recovery as well as in prepared foods; citric acid, which is used in the food and chemical industries (about 400,000 tons per year worldwide); polyhydroxybutyrate (PHB), a thermoplastic and biodegradable plastic that can be molded or made into fiber or sheets; antibiotics such as penicillins, cephalosporins, and tetracyclines; and many more products.

Fermentation offers the only economical route to the production of biopolymers like xanthan and other complex molecules such as the antibiotics and citric acid, and it will always be thus. In the case of other chemicals that can be produced from agricultural crops, crop residues, or processing byproducts by fermentation (ethanol, notably), the value of the crop as a chemical feedstock depends on the price of petroleum, the price of corn, and the cost of conversion.

Biotechnology can make, and has made to some extent, an impact in this area by increasing the efficiency of conversion and thus reducing raw material and production costs. Genetic engineering has been used to modify xanthan, but the properties of the new polymers are less desirable than those of the native product.

Enzymes

Enzymes are proteins, and genetic engineering provides the possibility of inserting an enzyme into an organism in which it was not originally present, removing an enzyme from an organism, or modifying an enzyme in an organism. So production of enzymes, which are also complex molecules that can be made only in living organisms, provides another interface with biotechnology.

Enzymes enjoy a wide range of applications—from use in cleaning products, to metal recovery processes, to leather treatment, to biotransformations such as modification of pharmacologically active compounds like steroids and prostaglandins. A current major use is in the production of corn syrup and high-fructose corn syrup (about 21.5 billion pounds annually) and maltodextrins. One enzyme produced by a genetically engineered microorganism, chymosin, is already used in commercial cheesemaking.

A recent advance in this area made possible by biotechnology is the use of heat-stable enzymes. Cyclodextrins, also made by enzymic conversion of starch, have now been introduced in the U.S. as an industrial product. Potential applications of cyclodextrins include controlled release of drugs, pesticides, flavors, and fragrances and their separation, purification, and protection.

We can expect that biotechnology will someday provide enzymes that convert starch into other new specific products, more heat-stable enzymes, enzymes that remove lignin from wood, enzymes that more effectively break down cellulose to glucose so that it can be used for alcohol production, enzymes that modify both simple and complex carbohydrate structures in specific ways, and enzymes that are used as processing aids.

Pharmaceuticals from Plants and Animals

It is possible to engineer plants to produce specific proteins and peptides. This allows high-value pharmaceutical peptides, including blood factors and growth hormones, to be obtained from plants. Other possibilities include the use of plants to make natural pesticides and other chemicals that can be extracted and purified. As of 1989, none of these value-added products were yet being produced commercially in this way.

Alternative uses for farm animals are currently being investigated. Because milk-producing animals are efficient bioreactors for converting feed into valuable proteins, it will be possible in the future to engineer animals to produce high-value pharmaceutical or therapeutic products in their milk. With the concern over AIDS and the safety of the blood supply, there is increasing need for the production of specific blood proteins to be used for transfusions.

Transgenic animals, engineered to produce these blood factors in their milk, could serve as a viable and safe alternative production system. Transgenic animals may serve as a more economic alternative than microbial fermentation for the large-scale production of human and animal vaccines, drugs, peptide hormones, monoclonal antibodies, and other high-value pharmaceutical products. These applications would clearly provide opportunities for a relatively small number of farmers. However, they are an important example of the potential opportunities that will be available in the agricultural sector as a result of biotechnological advances.

Socioeconomic Issues

Sustainable Agriculture

There is increasing concern that traditional agricultural practices, which rely heavily on the use of chemical pesticides, herbicides, and fertilizers, will compromise our ability to sustain a productive agricultural system in this country. In addition, there is increasing consumer demand for "organic" produce which has been grown in the absence of added agricultural chemicals.

Many of the applications of biotechnology to production agriculture are totally compatible with these sustainable agriculture goals. Genetic engineering of crops that naturally resist microbial and viral diseases and insect pests will decrease our dependence on chemical pesticides. Engineering of crops that tolerate more "environmentally friendly" herbicides will also have a positive impact on groundwater quality. Plants that are able to resist environmental challenges (saline soil, arid climates, extreme temperatures) will expand the amount of land available for food production.

Reducing production costs through lowering or eliminating requirements for pesticides, herbicides, and fertilizers should increase the efficiency of producing food and nonfood crops. And decreasing the use of agricultural chemicals will decrease or eliminate residues in food products and meet the recent consumer demand for residue-free food.

Biotechnology has the potential to increase agricultural production. This could potentially result in lower prices, which, in turn, could drive more farmers from production, lead to fewer and larger farms, and result in the decline of rural economies.

Vertical Integration in the Agricultural Sector

Biotechnology will be used to custom-design raw agricultural commodities with specific structural, functional, and processing characteristics, and it is important to consider what impact this would have on the traditional production agriculture sector, as well as the potential implications to the food- and nonfood-processing sectors. The availability of custom-designed crops could lead to increased vertical integration within the food and agricultural system. Much of the development of new varieties is being done by the private sector. Thus, new crops would undoubtedly be grown under contract, just as waxy maize and high-amylose corn now are.

Food processors would purchase less on commodity markets and engage in more contract growing of proprietary specialty crops. Processors would need to be willing to pay premium prices for specialty crops in order to encourage farmers to invest the time and effort to produce crops which would not be handled as bulk commodities. The amount that processors would be willing to pay for specialty ingredients would depend on the proprietary nature of the product, consumer demand for the product, and how much the consumer would be willing to pay for the perceived or real benefit.

Increased utilization of crops for nonfood applications would provide an additional outlet for agricultural commodities. It would also lessen our dependence on petroleum as a source of textiles, plastics, packaging materials, coating, etc., by providing an annually replenishable source of chemicals. The development of alternative crops such as kenaf or industrial rapeseed for nonfood uses would provide flexibility and additional markets for agricultural products.

Regardless of the nature of the crop, commodity, or specialty, farmers would be needed to grow and harvest crops. Farmers would have the option of contracting with end users of the specialty ingredients, unless market demand dictated what was to be grown. Other industries which service the food-processing sector for transportation and storage of raw agricultural commodities, intermediate processing of bulk ingredients, and packaging of raw or partially processed ingredients would also be affected by increased vertical integration.

Cost of Food

The use of custom-designed agricultural commodities or ingredients with consistent, predictable, and reproducible structural and functional properties would eliminate much of the variability, inefficiency, and waste associated with current manufacturing processes, thus decreasing the overall cost of processing the food. This, of course, would be balanced by the investment in research and development of these components; therefore, the real cost of food may not be decreased in the short run.

The improvement of enzyme systems for use in foods could expand the uses for raw agricultural products and for waste generated during food processing, creating valuable products from previously unusable material. This could decrease the cost of processing food, decrease the real cost of food for consumers, eliminate waste disposal problems, and create new products, industries, and employment.

The high-fructose corn syrup industry was virtually nonexistent until the advent of enzyme technology. This product now commands a substantial portion of the sweetener market. Cellulase enzymes could be used to convert undigestible cellulosic plant material to substrates which could be metabolized by animals or humans, or used as feedstock for microbial fermentations. Thus, biotechnology could influence the real cost of food and, at the same time, stimulate new business opportunities in the food ingredient and processing sectors.

Expansion of Markets

Many of the plants used for extraction of high-value ingredients are not native to this country and so must be imported. The ability to produce these compounds in tissue culture could decrease our dependence on foreign suppliers for these ingredients. For plants grown in other countries, tissue culture technology should not compete with traditional agricultural production. For those plants currently grown in this country, however, use of cell culture technologies might reduce the demand for certain crops and decrease the level of production at the farm level, unless alternative uses of the plant material could be developed.

The ability to alter genetic structure offers the potential to develop crops for industrial utilization; however, it is not likely to have much of an impact on the average farmer in the near future. The time lag before a new crop for industrial use could be developed via genetic engineering is considerable, and there is little incentive for increased industrial use as long as crude oil and natural gas prices are low. In other words, industrial utilization of our renewable resources is driven by economics rather than by social or national concerns. Nevertheless, research in this area should be intensified.

While we continue to apply new knowledge and technology to the improvement of plants and animals, the politics and economics of food still drive agriculture almost exclusively. But the time will come when the need for alternative sources of raw materials for nonfood industries will become an important factor in the equation. By then, perhaps, biotechnology will have developed to the point where it can help provide replenishable sources of the required organic compounds.

The use of biotechnology for construction of improved starter organisms, particularly if the focus is improvement of the nutritional quality of cultured products, could expand the use of milk, meat, vegetables, and cereal grains which serve as substrates for fermentative organisms. With the increasing ethnic populations, more accustomed to the inclusion of fermented foods in their diets, markets for these products could increase dramatically. If products were designed to meet a particular dietary requirement, such as the need to reduce cholesterol or calories, individuals who did not previously consume fermented foods might be interested in including them in their diets.

Currently, many fermented foods, especially exotic cheeses and premium wines, are imported from other countries. Improvement of the fermentation organisms might allow the manufacture of equivalent or superior products for use domestically and for export.

There are a number of advantages to the food industry for incorporating microbially derived ingredients in processed foods. There may be a marketing advantage, as consumers perceive that "natural" is better and may be willing to pay premium prices for natural products.

The improved consistency, availability, and predictability of processing of ingredients produced by microbial fermentation could improve the quality of products and decrease the cost of processing. Expanded use of agricultural commodities could provide alternative markets for farmers, and utilization of food processing waste streams could decrease the environmental impact and cost associated with safe disposal of waste. Unique ingredients would be appropriate for export to other countries, thus expanding markets for value-added byproducts of food processing.

There are many opportunities to expand markets by application of biotechnology. More resources should be devoted to developing alternative food and nonfood uses for raw agricultural products.

For example, BST would not be such a controversial issue if there were alternative uses for the anticipated milk surplus. There are many valuable proteins in milk which could be used for the production of pharmaceutical products, peptides, polymers, and specialty chemicals. In this way, the U.S. could take advantage of its tremendous capability to produce milk and, at the same time, expand domestic and international markets for milk-derived products. Producers, processors, and consumers would all benefit from such an approach.

Regulatory Climate

A potential obstacle to the broad application of biotechnology in agriculture and food processing is the regulatory climate. It is essential that the U.S.

Markets can be expanded by developing alternative food and nonfood uses for raw agricultural commodities.

develop a reasonable regulatory framework that will ensure the safety of foods derived from biotechnological processes yet not stifle innovation. The food and agricultural sectors are highly regulated, and without a well-defined regulatory structure, they will be very reluctant to invest the time and money developing products whose entrance into the marketplace might be delayed or prevented.

Other countries look to the U.S. regulatory system as a guide for developing their own systems, and it is important that the laws regulating biotechnology be consistent between countries or trade could be severely restricted. Also, in many cases, that of BST being one example, it will be impossible to determine if a product imported into this country has been produced using biotechnological techniques.

Consumer Perception

Consumers will need to embrace biotechnology if it is to survive and flourish in the agricultural and food sectors. If consumers perceive that a technology is not safe, it can have a dramatic effect on the ability or willingness of industry to use the technology.

A good example, meriting discussion at some length, is the controversy surrounding the use of BST. Surprisingly, BST has been, and continues to be, a very controversial product. This serves to illustrate how socioeconomic issues can affect the utilization of a technology which could prove to be a beneficial management tool for farmers. Although it is generally accepted that BST poses no hazard to humans or long-term negative effect on animals, many consumers have a very negative perception of BST. This is due, in part, to a general distrust of technology and a lack of understanding of the nature of BST and how it functions in the animal.

A primary concern appears to be the possible socioeconomic implications if BST were to be broadly adopted by farmers. Many question the rationale for producing more milk when there is a perceived surplus of this commodity. It has also been suggested that BST will further the decline of the family farm, because small farms may be less able than large farms to apply the technology.

BST has become a political issue in certain states, and legislation has been passed that temporarily bans the use of BST in Wisconsin and Minnesota. If BST were banned permanently in certain states, farmers in those states would be at a distinct disadvantage in the national marketplace. Similarly, if a national ban of BST were instituted in the U.S. but not in other countries, the ability of U.S. farmers to compete in the global marketplace would be compromised. This has led to a debate on whether or not economic justification should be included as a criterion for approval of a new product like BST. If so, it is important to consider who has the responsibility for making such a determination.

BST is just the first of many biotechnology-derived compounds to affect the food supply. The scientific community must be prepared to meet the unique challenges that each application may pose. For example, we will need to develop criteria to assess the safety of an engineered plant that produces its own insecticide and develop methods for communicating to consumers the relative risks and benefits of each new technology compared to traditional technologies. Industry, academia, and government agencies all share the responsibility of educating the public about biotechnology. With the decreasing science literacy in this country, this will become an increasing challenge.

And it is a challenge that ought not be ignored. As a market-driven industry, the food processing sector would be reluctant to process and market biotechnology-derived products if consumers were unwilling to purchase them. If sales would be compromised by the use of biotechnology-derived foods, they would not be used by the industry. With all of the food products available in the marketplace, processors know that consumers will exercise their option of purchasing a competitor's product if they have any reason to believe that a product is not safe. Farmers are also responsive to public perception and consumer demand and would be hesitant to embrace the technology if markets were compromised.

Who Decides?

Who decides when and how agricultural biotechnology should be used? Obviously, we all have a role to play in these decisions, and it is important that we work together to address the issues surrounding the choices we must make for the further development of biotechnology. It is critical that the public be made aware of what biotechnology is and how it could be used to meet some of the major challenges of the world: enabling agriculture and food processing to keep pace with population growth, controlling pollution of the environment and contamination of the water supply, and ensuring the health of the nation.

References and Additional Readings

Applewhite, T. H., ed. 1988. Proceedings of the world conference on biotechnology for the fats and oils industry. American Oil Chemists' Society, Champaign, IL.

Harlander, S. K., and T. P. Labuza, eds. 1986. Biotechnology in food processing. Noyes Publications, Park Ridge, NJ.

Morris, C.E. and A.E. Przybyla. 1989. Biotechnology blooms. Food Engineering. July, p. 51-58.

Regulatory considerations: Genetically-engineered plants. 1987. Summary of a workshop held at Boyce Thompson Institute for Plant Research, Cornell University. Center for Science Information, San Francisco, CA.

Taylor, S. E., I. Stith-Coleman, J. E. Mielke, and J. A. Johnson. 1989. Biotechnology: A public policy primer. Congressional Research Service Report for Congress.

CHAPTER FOUR

The application of biotechnology to food production and processing has implications for the safety and quality of our nation's food supply. To the extent that biotechnology lessens the use of chemical pesticides in production agriculture, chemical residues in foods should decrease. When plants are genetically engineered to enhance food quality, functionality, or processing or handling characteristics, appropriate measures must be taken to ensure that desirable attributes such as nutritional quality are not inadvertently diminished and that naturally occurring toxicants are not unknowingly increased. The safety of genetically engineered microorganisms used in food processing will need to be evaluated to ensure that they have not become pathogenic and that they do not produce any unwanted toxins or other harmful substances.

Biotechnology also has food safety implications in the area of detection of harmful substances. Conventional detection methods are often complex, labor intensive, and time consuming. Methods made possible through biotechnology are more rapid, more sensitive to the presence of microbes, and more specific in their ability to detect pathogens. Examples include monoclonal antibodies, used in immunoassay technology to detect foodborne bacteria and toxins, herbicide residues, and therapeutic agents, and DNA hybridization technology, which uses DNA probes to detect microbial pathogens. Although not reducing the foodborne illnesses caused by improper food preparation, biotechnological improvements in detection capabilities will likely increase the overall safety of our food supply.

4 Food Safety Issues in Biotechnology

Michael P. Doyle, University of Wisconsin-Madison
Elmer H. Marth, University of Wisconsin-Madison

Introduction

Application of biotechnology to food production and processing has the potential to increase yields of some raw foods over those obtained by conventional agricultural practices. Furthermore, possibilities exist to improve sensory, functional, and nutritional attributes of some finished products through biotechnological processes. However, these advances will have the anticipated impacts only if resulting foods are safe for human consumption.

This chapter deals with issues of food safety. First, there is a discussion of the status of food safety. This is followed by a summary of types of food biotechnology, which leads into concerns about food safety and quality. Societal implications of food biotechnology are considered in the concluding section of this chapter.

Status of Food Safety

The consumer expects food to be safe, thus that it will not cause any illness when it is eaten. In fact, since this is the common experience of the consumer, safety of food is generally assumed by most people in the U.S., except periodically when a problem occurs that receives considerable media attention. Sometimes food does become unsafe and causes illness. The most common reasons for lack of safety are (a) foodborne illnesses of microbial origin, (b) environmental contaminants in food, (c) toxicants naturally present in food, (d) additives, and (e) specific nutritional and dietary problems of certain consumers for whom some foods are never safe.

Foodborne Illnesses of Microbial Origin

Of the reasons just listed, foodborne illnesses of microbial origin are the most important in terms of people involved and economic losses. Epidemiologists have estimated that between 20 million and 80 million U.S. residents (i.e., up to one-third of the total population) annually experience at least one bout of food-associated, microbially induced diarrhea. Without even considering the other causes of unsafe food, it is evident that the economic loss (product recalls followed by destruction or reprocessing of product, loss of time by persons in the workforce, medical expenses, litigation, etc.) caused by foodborne illnesses of microbial origin is of major proportions.

Foodborne illnesses of microbial origin can be arranged into two groups: food intoxications and food infections, which includes in vivo toxin production by the causative agent. Major food intoxications are botulism, gastroenteritis caused by *Bacillus cereus*, mycotoxicoses, and staphylococcal poisoning. A food intoxication results when a toxigenic microorganism gets into food, grows and produces a toxin, and then the toxic food is consumed. Some additional information about the common food intoxications is summarized in Table 1.

A food infection results when a microorganism, most commonly a bacterium, virus, or parasite, gets into food and then food with the viable microorganism is consumed, whereupon illness results. Examples of diseases resulting from food infections include the relatively common salmonellosis and the relatively uncommon Q-fever. Tables 2, 3, and 4 summarize additional information on foodborne infections caused by bacteria, viruses, and parasites, respectively.

The authors wish to thank Dr. W. J. Brill, Department of Bacteriology, University of Wisconsin-Madison, and Dr. J.R. Steele, Department of Food Science, University of Wisconsin-Madison, for providing some of the source materials used to prepare this chapter.

Table 1. *Common Forms of Food Intoxication Caused by Microorganisms.*

Disease	Causative agent	Examples of foods involved*
Botulism	*Clostridium botulinum* (neurotoxin)	Canned low-acid foods (usually home canned), smoked fish, cooked potatoes, frozen pot pies, meat loaf, stew, fermented fish eggs, fish
Gastroenteritis	*Bacillus cereus* (enterotoxins)	Boiled or fried rice, cereal products, soups, custards, meat loaf, sausage, cooked vegetables, refried beans
Mycotoxicoses	Numerous molds, largely in the genera *Aspergillus, Penicillium, Fusarium, Alternaria, Claviceps* (various toxins are involved)	Peanuts, corn, cereal grains, cottonseed, pecans, Brazil nuts, sweet potato, cassava, certain animal products from animals that consumed toxic feed
Staphylococcal poisoning	*Staphylococcus aureus* (enterotoxins)	Ham, other cooked meat and poultry products, cream-filled pastries, cheese, whipped butter, dry milk, meat salads, sauces, gravies, hard-boiled eggs

* *Control involves use of proper hygienic measures to minimize contamination plus suitable means to keep the organism from growing in foods. Once formed, ordinary heating of foods will not inactivate mycotoxins and staphylococcal enterotoxins, whereas the others may be inactivated if heating is sufficient.*

Environmental Contaminants

Food plants grow in soil, often are irrigated, commonly are treated to control pests, and always are handled during and after their harvest. Such exposure to the environment affords ample opportunity for contamination by microorganisms and also by chemicals of the types listed in Table 5. An example of an environmental contaminant is the accumulation of mercury in tissue of mushrooms grown on soil containing more than the normal amount of mercury. This can happen when mercury-laden wastes are discharged onto the soil.

Natural Toxicants

Some plants are so constituted that they synthesize toxicants during growth, and these substances accumulate in tissues of the plants. Depending on the toxicant, variety of plant, and amount consumed, such toxicants can cause illness or even death. Examples include oxalates in rhubarb, glycoalkaloids in potatoes (especially in areas of tubers exposed to light), and cyanide compounds in lima beans. Molds commonly infest many raw agricultural commodities. Over 80 species of molds, if allowed to grow, can produce secondary metabolites (mycotoxins) that are hazardous to humans and animals. Additional examples of natural toxicants in food are listed in Table 5.

Additives

There are over 2,000 additives acceptable for use in foods. Included are such direct additives as antimicrobial agents, antioxidants, colorants, flavors, stabilizers, thickeners, leavening agents, acidity control agents, emulsifiers, sweeteners, enzymes, and certain microorganisms used to make fermented foods. Indirect additives include certain solvents, lubricants, release agents, and packaging components. Although not considered as indirect additives, animal drugs such as antibiotics used to treat illnesses can contaminate some foods of animal origin.

Although numerous additives can occur in foods either indirectly or through direct addition, 95 percent of the amount of consumed additives consists of sweeteners, salt, starch, mustard, sodium bicarbonate, yeast, caramel, citric acid, carbon dioxide, and black pepper.

That some additives sometimes can be unsafe for certain consumers was demonstrated recently for sulfites. These widely used preservatives (in wines and certain fruit and vegetable products) can cause a major allergic-type response in susceptible persons, so use of sulfites has been restricted.

Nutritional and Dietary Problems

Some foods that are safe for most persons are unsafe for others. An example is foods containing phenylalanine, which are unsafe for persons suffering from phenylketonuria. Other examples of conditions that make consumption of certain foods problematic for some persons include obesity (too much food), cardiovascular disease (certain foods of animal origin), cancer (fats, cured meats and others), diabetes (high sugar foods), and malnutrition (too little food or specific nutrients).

Types of Food Biotechnology

Food biotechnology is broad in scope and thus potentially can have numerous applications in food production and processing. Included in food biotechnology is a collection of processes that involve use of biological systems occurring in animals, plants, and microorganisms. The scientific base for food biotechnology comes from disciplines such as genetics, biochemistry, cell biology, and microbiology, whereas the applications base comes from food science, chemical engineering, and instrumentation technology.

Animal Biotechnology

Foods of animal origin, e.g., meat, poultry, milk, eggs, and seafood, are a major component in the diet of most people. Since there is a large demand for such foods, techniques of biotechnology have been applied to enhance feed efficiency and growth rates of food-producing animals. An example of such an application is the use of bovine somatotropin (BST) to enhance milk production through more efficient use of feed by dairy cows. This has become possible because the cow's gene for producing the peptide hormone BST has been cloned into a bacterium and thus a large quantity of the hitherto scarce hormone has become available.

A new technique associated with an improvement in animal breeding is calf cloning. In this form of biotechnology, an embryo early in its development is disrupted to yield individual cells, each

Table 2. *Common Forms of Food Infection or In Vivo Toxin Formation Caused by Microorganisms.*

Disease	Causative agent	Examples of foods involved*
Brucellosis	*Brucella* spp.	Raw milk, goat cheese
Cholera	*Vibrio cholerae* (enterotoxin)	Raw fish, shellfish and crustacea, water, foods prepared with contaminated water
Diarrhea	*Escherichia coli* (enteropathogenic types)	Soft cheese, ground beef, raw milk
Enteritis	*Clostridium perfringens* (enterotoxins)	Cooked meat, poultry, gravy, sauces, soups, refried beans
Gastroenteritis	*Vibrio parahaemolyticus*	Marine fish, shellfish, crustacea
Listeriosis	*Listeria monocytogenes*	Vegetable products, raw milk, soft cheeses, meat, poultry products
Pharyngitis	*Streptococcus* sp. (beta-hemolytic)	Raw milk, salads
Q-Fever	*Coxiella burnetii* (a rickettsia)	Raw milk
Salmonellosis	*Salmonella* spp.	Poultry and meat products, eggs and egg products, raw milk, milk products
Shigellosis	*Shigella* spp.	Salads, vegetables, water
Yersiniosis	*Yersinia enterocolitica*	Milk, tofu, water

* *Control commonly involves use of suitable hygienic measures to minimize contamination of food together with proper processing to kill the pathogens before food is consumed. Care is needed to prevent recontamination of processed foods.*

of which is then placed into the uterus of a receptive cow and permitted to develop. In this way up to 32 identical offspring can result from a single fertilization event. The potential impact of this development on production of foods of animal origin is incalculable.

Plant Biotechnology

Cereals and other plants are consumed by virtually everyone worldwide. The diet of many humans, either by choice or because of supply problems, consists entirely of plant-derived foods. Techniques of biotechnology have been employed to improve resistance of plants to stress (e.g., freezing), invasion by insects, and viral diseases. These techniques also have been used to reduce dependence of plants on nitrogen-containing chemical fertilizers.

For millennia, the root-nodule bacteria have been associated symbiotically with leguminous plants. In this association the bacteria provide the plants with nitrogen obtained from the air. Other nonsymbiotic, nitrogen-fixing bacteria also exist in the soil. Attempts are being made to transfer this ability to fix nitrogen to plants not traditionally associated with the process. Furthermore, the functionality (e.g., foaming characteristics, bulk-density) and nutritional value (e.g., raising the level of amino acids that are currently deficient in corn protein) of plant foods can be improved through application of biotechnology.

Microbial Biotechnology

Some raw materials of animal or plant origin are converted to products suitable for consumption through action of microorganisms, often the lactic acid bacteria. Examples include certain sausages, cheeses, and sauerkraut. This important group of bacteria is susceptible to infection by bacteriophages, so the ability to convert sugar into lactic acid is lost. Biotechnology has been used to construct bacteriophage-resistant strains of these bacteria. The lactic acid bacteria together with related kinds, through activity of their enzymes, contribute to developing sought-after flavor, aroma, and texture in some foods. This commonly is a slow process with variable results; standardization and acceleration of the process should be possible through application of biotechnology.

"Probiotics" is a term applied when microorganisms are consumed because they are thought to be beneficial for human health. Examples of such bacteria are *Lactobacillus acidophilus* and *Bifidobacterium bifidum*. Their ingestion is advocated by some as a way to reduce the incidence of colon cancer and to reduce levels of serum cholesterol. Although clinical data so far are incomplete and inconclusive, if they ultimately show that expected benefits do occur, it may be possible to modify the bacteria and enhance their positive impact on human health.

Use of microorganisms to produce food ingredients, food flavors, and food preservatives from food wastes or other low-cost substrates offers further opportunities for application of biotechnology.

Worldwide, the food industry uses more than 50 percent of commercially produced enzymes. Some examples are amylase to hydrolyze starch, invertase to produce invert sugar, and chymosin to coagulate milk. Biotechnology is being used to improve yield, enhance stability, and alter specificity of enzymes.

Rennet extract, obtained from the true stomach of young calves, contains chymosin, the enzyme responsible for coagulating milk in the cheesemaking

Table 3. *Common Forms of Disease Caused by Foodborne Viruses.*

Disease	Causative agent	Examples of foods involved*
Gastroenteritis	Norwalk-like viruses	Clams, oysters, cockles, green salads, pastry, frosting
Gastroenteritis	Enteric viruses (ECHO, rota-, and polioviruses)	Coleslaw, water
Hepatitis A (infectious hepatitis)	Hepatitis A virus	Raw shellfish, salads, cold cuts, fruit, water

* *Control involves avoidance of contaminated raw foods, proper cooking of foods likely to be contaminated, and preventing recontamination of cooked foods.*

process. Increases in cheese production and decreases in the calf population have caused a shortage of available chymosin. To overcome this problem, the calf chymosin gene has been synthesized and introduced into a microbial cell. The microbial cell then produces a high level of chymosin that is readily available and, when suitably processed, is cheaper and purer than the product obtained from calves.

Food Safety and Quality Concerns

Biotechnology in food production and processing may directly and indirectly affect the safety and quality of food. In some instances the effect may be beneficial, in others detrimental. For example, vegetables and other crops may be genetically engineered to introduce resistance to pests or to eliminate an enzyme that causes deterioration of the food, yet care must be taken to avoid decreasing nutritional quality, affecting expression of naturally occurring preservatives, or increasing production of naturally occurring toxicants.

Genetically engineered microorganisms can be designed to produce desirable enzymes or metabolites for use in food production or processing, but it is important that these genetic changes do not create organisms that may be harmful to human and animal health through toxin production or acquisition of properties (virulence) that induce illness. Unique and unconventional processes may be developed to integrate products of bioengineered microorganisms into food processing. Such processes should be designed to eliminate or control foodborne pathogens that may be present in ingredients or the environment.

An important contribution of biotechnology to improving food safety is use of techniques (i.e., gene probes and monoclonal antibodies) of this branch of science for more rapid, and, in some instances, more sensitive detection of foodborne pathogens, toxins, and residues than is possible using conventional methods. This application of biotechnology and others are discussed later in this chapter.

Genetically Engineered Vegetables and Crops

Pest-Resistant Varieties

Widespread use of pesticides has led to contamination of drinking water and underground water supplies in several

Table 4. *Common Foodborne Parasitic Diseases.*

Disease	Causative agent	Example of foods involved*
Amebiasis (amebic dysentery)	*Entameba histolytica*	Fruits and vegetables
Angiostrongyliasis (eosinophilic meningoencephalitis)	*Angiostrongylus cantonensis*	Crabs, slugs, prawns, shrimp, snails
Anisakiasis	*Anisakis* sp., *Phocanema* sp., and *Porrocaecum* sp.	Rock fish, herring, cod, squid
Diphyllobothriasis	*Diphyllobothrium latum* (fish tapeworm)	Perch, pike, turbot, salmon
Giardiasis	*Giardia lamblia*	Salmon, water, fruits and vegetables
Taeniasis	*Taenia saginata* (beef tapeworm) *Taenia solium* pork tapeworm)	Beef, pork
Toxoplasmosis	*Toxoplasma gondii*	Meats
Trichinosis	*Trichinella spiralis*	Pork, bear meat, walrus flesh

**Control commonly involves use of proper hygienic measures and adequate cooking of the foods that can carry the parasites.*

Table 5. *Environmental Contaminants and Natural Toxicants that Can Cause Unsafe Food.*

Environmental contaminants	**Examples of foods involved**
Heavy metals (Hg, Pb, Cd, As)	Numerous plant products
Polychlorinated biphenyls	Fish
Polybrominated biphenyls	Rare
Pesticides	Numerous plant products
Herbicides (dioxin)	Numerous plant products
Fungicides	Numerous plant products
Radioactivity	Plant products, milk
Shellfish poisons	Shellfish
Natural toxicants	
Oxalates	Rhubarb
Glycoalkaloids	Potatoes and other plant products
Cyanide compounds	Lima beans (some varieties)
Mushroom toxins	Mushrooms
Vasoactive amines	Cheese, wine, certain fish
Psychoactive substances	Certain spices
Allergens	Milk, wheat, fish
Mycotoxins*	Numerous plant and animal products

* *See Table 1.*

locations throughout the U.S. Other consequences of excessive use of these chemicals include presence of residues in the food chain, development of minor pests into major ones through reduction of natural enemies, and propagation of insecticide-resistant strains of pest species. By genetically engineering plants for resistance to disease and insects, the need for pesticides can be dramatically reduced if not eliminated.

The bacterium *Bacillus thuringiensis* (*Bt*) produces an insecticidal substance that kills many noxious pests, such as caterpillars, beetles, and moths. Preparations of this organism have been used safely worldwide as microbial pesticides for more than 20 years. Such preparations are considered to be extremely safe because they commonly are harmless to humans, animals, and useful insects such as bees, and because they are highly specific, being toxic to only certain types of insects.

Molecular biologists have isolated and characterized the genes that encode the *Bt* toxin and have introduced these genes into plants to confer insect resistance. Tomato plants with *Bt* toxin gene are now being evaluated in field tests for pest resistance.

United States Department of Agriculture statistics indicate that farmers spend $150 million annually on chemical insecticides, yet caterpillar-type insects alone cause $450 million damage a year. Introduction of bioengineered pest-resistant plants could save agriculturists millions of dollars annually and protect the environment and food chain from future contamination by pesticides.

However, a potential concern is that widespread use of bacterial pesticides will cause insects to develop resistance to "natural" insecticidal agents more quickly. Another consideration that has yet to be addressed is whether consuming substantial concentrations of *Bt* toxin that is found throughout plant tissue is as innocuous as naturally occurring *Bt* toxin that is produced and deposited in a crystalline form by bacteria.

Herbicide-Resistant Varieties

In contrast to using biotechnology to reduce or eliminate pesticide application by developing plants with self-producing insecticidal agents, scientists also are using biotechnology to develop herbicide-resistant plants. Tobacco plants resistant to atrazine have been developed and field-tested. Genetic resistance in crops to several other herbicides also is being sought. Although such resistance is potentially beneficial for crop production, this line of research could lead to greater and extended use of herbicides and thus could result in further serious contamination of water and the environment.

Changes in Nutritional Quality

There is great interest in using biotechnology to modify plants to enhance desirable characteristics that may benefit the food processor and ultimately the consumer. Examples include increasing the solids content of tomatoes to improve yields of catsup and tomato paste; eliminating the lipoxygenase gene from sweet corn and other vegetables to avoid development of off-flavors during frozen storage and thus eliminate the need for blanching; and producing a more flavorful popcorn that will be readily consumed without salt and butter.

However, in developing vegetables and other crops with desirable processing characteristics, care must be taken to maintain the nutritional integrity of the food. It has been shown in classical plant breeding studies that some tomato varieties bred for mechanical harvesting had 15 percent less vitamin C than did their predecessors. Unfortunately, in selecting genes that made tomatoes desirable for mechanical harvesting, genes needed for maintaining high levels of vitamin C were eliminated. It is especially important to avoid an appreciable reduction of those specific nutrients in plants that are commonly recognized as good sources of such nutrients.

Biotechnologists can develop tomatoes with desirable processing characteristics, but this must not be accomplished at the expense of nutritional integrity.

From another perspective, considerable research is being done to improve the nutritional quality of crops through biotechnology. This is particularly important in developing countries where certain crops with limited nutrients are staples in the diet. Through genetic engineering, attempts are being made to increase the protein and decrease the cyanide content of tropical root crops such as cassava and taro. Research is being done to develop a variety of rice that produces carotene to help prevent vitamin A deficiency in countries where rice is the principal component of the diet. Studies also are being done to increase the nutritional quality of protein in the Russett Burbank potato and in soybeans.

Changes in Naturally Occurring Preservatives

Many plants, including their seeds and fruits, contain antimicrobial substances which inhibit growth of spoilage microorganisms and/or human pathogens. Cranberries contain benzoic acid, which inhibits mold growth. Wheat endosperm and other cereal grains contain purothionins, which inhibit some human pathogens and yeasts. Anthocyanin pigments present in many fruits and cocoa beans inhibit several types of bacteria. Tomatoes contain α-tomatine, which inhibits fungi and bacteria; they also contain citric acid, usually at levels sufficient to inhibit *Clostridium botulinum*, the cause of botulism.

Consideration should be given to changes that may occur in the expression of these natural preservatives when plant genes are manipulated for strain improvement. Fruits and vegetables with reduced levels of preservatives may spoil more rapidly from microbial degradation than those with normal levels of natural microbial inhibitors.

Additionally, consideration should be given to food safety concerns. Traditionally, tomatoes that have been produced contained enough citric acid to prevent production of the deadly botulinal toxin in minimally processed canned products. This allowed the homemaker to give tomatoes a mild heat treatment during canning for preservation. However, developing tomatoes with less acidity would minimize their naturally occurring antibotulinal properties and could result in outbreaks of botulism if the homemaker used the traditional canning process.

Changes in Naturally Occurring Toxicants

Certain vegetables and other crops contain naturally occurring toxicants at levels that normally would not produce a toxic response when eaten. Progoitrin is present in cabbage, kale, brussels sprouts, cauliflower, broccoli, kohlrabi, turnips, rutabaga, and rape and crambe seeds. Goitrin, which inhibits thyroid activity and results in goiter, is a toxic breakdown product of progoitrin. Linamarin, a cyanogen that yields deadly hydrogen cyanide upon treatment with acid or appropriate hydrolytic enzymes, is present in cassava, linen flax, and lima beans. Glycoalkaloids, such as solanine, are naturally occurring toxicants in potatoes, peppers, tomatoes, and eggplants. When ingested at high concentrations, glycoalkaloids can cause severe intestinal disorders and even death.

Thus, we must be sure that genetic engineering does not inadvertently turn on the genes that regulate production of such toxicants which are naturally formed at low levels, thereby resulting in hyperproduction of these toxins. Furthermore, it also may be possible through introduction of genes from other plant species to inadvertently introduce new toxins into hitherto toxin-free agricultural crops.

Research with classical genetics of plants inadvertently has resulted in development of varieties with improved qualities but also increased levels of naturally occurring toxins. A classic example is development of a new variety of potato named "Lenape," which was valued for making potato chips. It was later discovered, after an individual became ill from eating some Lenape potatoes, that they contained about twice the amount of glycoalkaloids found in commercial varieties of potatoes.

With gene transfer moving at a faster pace through biotechnology than through classical plant breeding, the likelihood is increased of changing toxin levels or introducing totally new ones in plants. New cultivars developed through bioengineering should be evaluated for expression of relevant toxins before being commercialized.

Products of Genetically Engineered Microorganisms

Considerable research is underway to apply genetic engineering to microorganisms for production of beneficial metabolites that may be used as food ingredients or aids in food processing. Microorganisms may be genetically modified to produce large quantities of flavoring agents (e.g., diacetyl, which has a buttery flavor; citronellol, a fruity flavor; methylbutanol, a malty flavor; tetramethylpyrazine, a nutty flavor), fragrances (e.g., fatty acid esters, which smell fruity; gamma-decalactone, a peach odor; geraniol, a roselike odor; 3-methoxy-3-isopropylpyrazine, a potato odor; 3-methylbutylacetate, a banana odor), amino acids, vitamins, acidulants, thickeners, sweeteners, and enzymes (e.g., chymosin used in cheese production, amylases used in brewing, invertase used in candy making, glucose isomerase used in high-fructose corn syrup production). Because many microorganisms may produce toxins or harmful effects in humans, consideration must be given not only to the beneficial metabolites produced by the organism but also to the safety of the organism used to produce them.

Several points should be considered in evaluating the safety of genetically engineered microorganisms and their metabolites for use in foods. For example, it is important to fully characterize the genetic alterations that are introduced to establish that the gene sequences do not code for biosynthesis of harmful substances. Information is also needed to show that the microorganism is not pathogenic and does not produce antibiotics or toxins. It is important to demonstrate that the

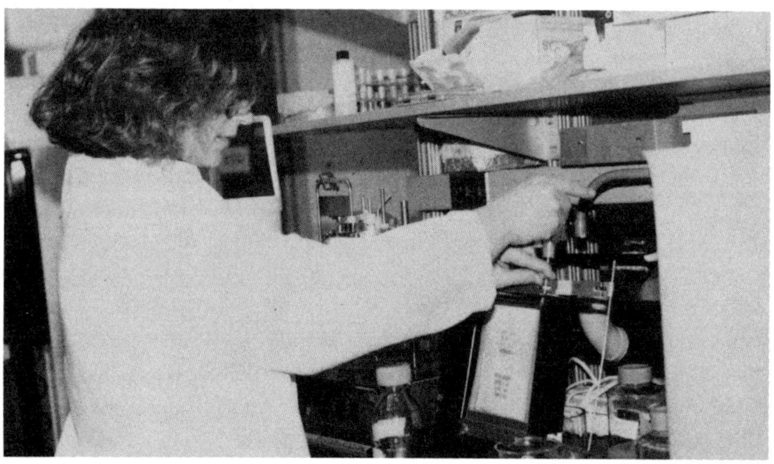

Traditional methods of monitoring foods and food ingredients for contaminants are often complex, slow, and labor intensive, whereas biotechnology can and is being used to develop more rapid and sensitive methods.

modified microorganism is a pure culture or that the end product is free of viable organisms and biologically active DNA that encodes resistance to antibiotics that are in common clinical use. The final food product should be of food-grade quality and free of microbial toxins.

There is no reason to believe that the relatively new capability to genetically engineer microorganisms presents any unique hazards that are clearly different from those just mentioned. This is supported by a recent report of the National Academy of Sciences, which concluded that "the risks associated with introduction of recombinant DNA-engineered organisms are the same in kind as those associated with introduction of unmodified organisms or organisms modified by other methods." Prudence, however, dictates that a thorough evaluation of the safety of the genetically modified microorganism and its end product(s) should be completed before they are used in food.

Processing of Food-Use Products of Genetically Engineered Microorganisms

Clearly, it is critical that products of genetically engineered microorganisms be free of harmful substances. Of equal importance, treatment of these products, including recovery, clarification, purification, packaging, and storage, should be done under hygienic conditions to prevent contamination with and growth of microbial pathogens and spoilage microorganisms. Processing of these products should be done using the same Good Manufacturing Practices that are required by law and applied by the food industry.

Improved Methods for Detecting Harmful Bacteria, Toxins, and Residues in Foods

A major priority of the food industry is to ensure a safe and wholesome food supply free of pathogens and harmful substances. A great commitment to quality control is required to accomplish this. Part of an effective quality-control program involves monitoring ingredients and finished products for hazardous contaminants. Many traditional methods used for this purpose are complex, labor-intensive, and time consuming. Methods that are more rapid and sensitive and that are based on monoclonal antibodies and DNA hybridization can be developed through biotechnology.

Immunoassays, using monoclonal antibodies, can be developed to detect any substance for which a specific antibody can be produced. Examples include microbial pathogens, toxins, hormones, pesticides, and adulterants. The enzyme-linked immunosorbent assay (ELISA) is the most common immunoassay in use. It can be rapid, accurate, sensitive, and relatively simple to do. Monoclonal antibodies often are more advantageous than traditional antibodies for such applications because they can be produced in large quantities from a stable, secreting hybridoma line that will grow continuously in tissue culture media or can be held frozen indefinitely and recalled for later use. Furthermore, monoclonal antibodies from a specific hybridoma are of a defined specificity and of a uniform affinity and, if appropriately selected, can be highly specific for a bacterium, toxin, hormone, or adulterant. Presently, monoclonal antibody-based ELISA procedures are available to detect harmful foodborne bacteria such as *Salmonella*, foodborne toxins such as mycotoxins and staphylococcal enterotoxin, herbicides such as atrazine, and therapeutic agents such as sulfamethazine.

The more rapid and sensitive methods of detecting food pathogens made possible through biotechnology will result in food that is both cheaper and safer.

Another application of biotechnology for rapid detection of foodborne pathogens is use of DNA hybridization technology. This technology potentially allows detection of any microorganism of interest in foods.

DNA probes are specific, single-stranded nucleic acid sequences that encode for markers, such as virulence factors, that are specific to microorganisms of interest. These probes are labeled, often with a radioactive marker (^{32}P), so they can be tracked. When the probes are introduced into a clinical specimen or food enrichment culture, they specifically hybridize or bind to a complementary DNA sequence on another strand of DNA that is present in the microorganism of interest.

The hybridization event indicates that the microorganism of interest is present in the specimen, and hybridization of the DNA is detected by assaying for the label of the probe. One measures radioactivity if ^{32}P is used as the label. Presently, DNA probe technology is available to detect a variety of microbial pathogens, including *Salmonella*, vibrios, *Listeria*, *Campylobacter jejuni*, and *Shigella*. Several testing laboratories have adopted this technology for daily routine screening of numerous food samples.

Societal Implications

Implications of food biotechnology for society are at least threefold: benefits, problems, and perceived but not real (from a scientific viewpoint) problems.

Benefits

As mentioned earlier, application of biotechnology can lead to laboratory methods which are likely to be more rapid and sensitive than existing methods for detection of pathogens in food. Additional methods might be developed for spoilage microorganisms. Use of the methods could lead to economies in laboratory operations and in storage of products (early release from storage because information from tests is available sooner than when traditional methods are used). This can lead to less expensive food for the consumer or at least a smaller increase in price over time than might have occured if conventional tests were used.

Another possible benefit is production of crops with minimal use of fertilizers (through increased use of nitrogen fixation) and pesticides (through development and use of pest-resistant varieties of plants). Such applications could lead to reduced (from present minimal levels) levels of chemical residues in foods. Many consumers would view this as beneficial.

Use of biotechnology to develop improved strains of bacterial cultures for use in food fermentations, to increase yields of enzymes, and to synthesize standardized food ingredients (e.g., flavors) in abundance, can lead to products that are uniform from lot to lot in such attributes as texture, flavor, aroma, and color. Many consumers would appreciate such consistency in the foods they purchase.

Problems

Real problems created by application of food biotechnology are difficult to foresee, although some possible problems have been mentioned earlier in this chapter. One problem not mentioned earlier might be the need to recall products which tested positive for a pathogen using techniques based on biotechnology but which would have tested negative by conventional methods. Although this might be a problem for the industry involved, it also can be viewed as a benefit to the consumer.

Increasing the content of naturally occurring toxicants in plants through biotechnological manipulations to improve other characteristics might be a problem and needs to be considered before modified plants are marketed as foods.

Another problem is that of developing suitable and meaningful regulatory controls for products derived through use of biotechnological techniques. What tests need to be done? Will this vary from product to product? If so, how? Will real and perceived concerns of consumers be satisfied by testing and use of associated control measures?

Use of biotechnology to produce more food that is safer and has better keeping quality than food currently available may contribute to solving the problem of food shortages in some parts of the world. But will it?

Perhaps it will provide a short-term, humanitarian, partial solution. However, biotechnology is unlikely to be a long-term solution and may create more problems than are solved. More safe food leads to ever more burgeoning populations, which then will again require more food. Furthermore, such burgeoning populations will put further demands on various resources and will create ever increasing amounts of solid waste and pollutants in water and air, problems for which food biotechnology will not provide solutions. Unless an increase in the food supply is coupled with responsible human reproduction, problems of the sort just mentioned will grow.

Except for instances mentioned earlier in this chapter, food biotechnology probably will have little or no impact on the incidence of foodborne illness caused by microorganisms. These problems result primarily because food is handled improperly.

Perceived But Not Real Problems

At times the public's confidence in science as the means to ensure safe food is badly shaken. Is cholesterol in the diet good or bad? What about sodium? Fiber? Polyunsaturated versus more saturated fats? *Salmonella* in pasteurized milk? *Salmonella* in cheese? Other examples can be cited. These and other problems may have occurred in the "good old days," but the science of those days was unable to recognize the cause of the difficulties. Hence, the public was uninformed and happy. This is no longer true, and thus, to some extent, science has become its own worst enemy.

In addition to the concern just mentioned, biotechnology suffers because it is a means to modify the cell's genetic material, which many hold as inviolate. Humans are attempting to "play God" and, even worse, are doing so to benefit the "big greedy companies." This is the image that has been created by opponents of biotechnology, and they have been aided by the media reporting the news.

Unfortunately and far too often, fact has not been sorted from fiction. This is clearly evident from the situation with bovine somatotropin (BST) in Wisconsin and elsewhere. The real issues are clouded by rhetoric about greedy companies and saving the family farm, by politicians using an issue to suitably position themselves for the next election, and by use of scare tactics so consumers do not buy milk from BST-treated cows and thus food processors refuse to handle the milk.

Although the problems are perceived rather than real, they become real when perception reaches the magnitude just described. Many of the arguments we hear today once were leveled at other technological advances—pasteurization of milk and artificial insemination, to name two such advances. Thus, if history offers a basis for judgement, we can be optimistic about the future of food biotechnology.

References and Additional Readings

Committee on Communicable Diseases Affecting Man. 1987. Procedures to investigate foodborne illness. International Association of Milk, Food and Environmental Sanitarians, Ames, IA.

Davis, B., and D. L. Ritter. 1989. How genetic engineering got a bad name. Imprimis 18(2):1-5.

Doyle, J. 1988. Potential food safety problems related to new uses of biotechnology. p. 49-73. In Biotechnology and the food supply. National Academy Press, Washington, DC.

Doyle, M. P., ed. 1989. Foodborne bacterial pathogens. Marcel Dekker, Inc., New York, NY.

Fischoff, D. A. 1987. Role of biotechnology in pesticide development: Bacillus thuringiensis as an example. p. 241-252. In L. G. Costa, et al. (eds.) Toxicology of pesticides: Experimental, clinical and regulatory aspects. Springer-Verlag, Berlin.

National Academy of Sciences. 1987. Introduction of recombinant DNA-engineered organisms into the environment: Key issues. National Academy Press, Washington, DC.

Roberts, H. R., ed. 1981. Food safety. John Wiley and Sons, Inc., New York, NY.

Subcommittee on Naturally Occurring Toxicants in Food. 1973. Toxicants occurring naturally in food. 2nd ed. National Academy of Sciences, Washington, DC.

Wasserman, B. P., T. J. Montville, and E. L. Korwek. 1988. Food biotechnology. Institute of Food Technologists, Chicago, IL.

CHAPTER FIVE

In this chapter the environmental risks and benefits associated with agricultural biotechnology are discussed from the biological, the public policy, and the social-psychological perspectives. These perspectives reflect the complex debate surrounding the environmental implications of biotechnology.

In recent decades there has been a shift in the biological sciences to the current paradigm that organisms exist within a biological continuum. Consequently, risk and benefit analysis, based on biological data, also has changed. Risks of agricultural biotechnology can be grouped on a time continuum, ranging from short-term to intermediate to long-term effects. As we move further across the continuum into the long-term effects, the lack of adequate information makes it increasingly difficult to assess and manage risks. Because of this uncertainty concerning risks, the public is increasingly insisting that the government assess risks and develop risk-management strategies for agricultural biotechnology.

Public policies to deal with environmental risks occur in two forms: ex ante *regulation and* ex post *judicial law. Socially optimal environmental policy requires both these forms. The lack of reliable information on the likely outcome of new biotechnologies hampers the development of appropriate environmental policies. Even if it were possible to generate more information, from a social-psychological perspective information is subject to different interpretations based upon society's value orientation. This can lead to uncertainty, conflict, and outrage. Only by providing adequate, reliable, acceptable, and appropriate information to all parties throughout the process of developing biotechnology can the cost of risk be minimized and the benefits maximized.*

5 Environmental Risks and Benefits of Agricultural Biotechnology

James J. Marois, University of California, Davis
James I. Grieshop, University of California, Davis
L. J. (Bees) Butler, University of California, Davis

This chapter is concerned with environmental risks and benefits associated with agricultural biotechnology. The environmental risks detailed are discussed from three distinct perspectives: the biological, the public policy, and the social-psychological. This approach is essential for an informed consideration and debate of the risks and benefits of agricultural biotechnology. While some observers may be concerned with the technology and risk assessment and management that stress scientific information and points of view, others may also incorporate economic, social, psychological, cultural, and political information and considerations.

This chapter also presents some methods as to how risks, especially the environmental/biological, might be managed and regulated, as well as how the debate between conflicting perspectives might be managed. This chapter provides background for the chapter on regulatory policy.

In this chapter, certain terms and concepts are used that require statements of definition. These include risk, benefit, risk assessment, and risk management.

Risk is defined as the probability of encountering "negatively valued events" as a consequence of some act. Risk is incurred when the probability of hazard, peril, or loss is more than zero. Risk implies uncertainty in relation to outcome. In the case of agricultural biotechnology, potential risks exist with the development and/or application of products or practices. In an environmental perspective, agricultural biotechnology risk usually refers to observable hazards (e.g., accidental release of a transgenic organism) that may be viewed as probable, harmful, and unwanted.

Benefits occur as the result of the application of a technology that helps solve a problem or increase agricultural productivity. Risk and benefit, particularly in an academic sense, do not fit together well. Whereas risk tends to be probabilistic, benefits are not.

Risk assessment by regulatory agencies is a process designed to systematically consider, document, and estimate impacts on human health and the environment. Hazards are identified, the nature and extent of exposure are estimated, and the probability of harm is calculated. A classic outcome of risk assessment is the EIR, or Environmental Impact Report. Risk assessment is designed to estimate the probability of various outcomes related to some action.

Risk management is concerned with various hazard-reducing or risk-reducing actions and with the enforcement of actions believed to reduce identified risks to acceptable levels. Examples of risk-reducing actions may include controls on research, manufacturing, and product use, and monitoring of biotechnology procedures. Although regulation may be the domain of state and/or federal agencies, efforts to create unified approaches for managing risks through regulation have yet to be satisfactorily developed. Society has increasingly insisted that government assess risks and develop effective management schemes for a host of technologies, including agricultural biotechnology.

In effect, society has demanded that more information be readily available to guide conclusions and decisions relative to agricultural biotechnology. Clearly there is a need for information for purposes of risk assessment and risk management.

Since information is a central limiting factor, two points need to be made about it. First, information on environmental risks is inadequate. Second, even when two parties look at the same information, different interpretations as to the meaning can result. Thus, factors other than the factual nature of the information are operative. In particular, social-psychological factors, such as motivation, prior experience and learning, perception, culture, and values, operate to influence interpretations. In both cases, lack of information and different interpretations of the same information often result in uncertainty and controversy.

Biotechnologies in agriculture, as they are currently understood, do pose potential environmental risks. Some are known; others are not. Some are controllable, with others the controllability is unknown or uncertain. Some individuals view such risks as knowable and manageable; others believe such risks are too complex for our understanding. In fact, no technology, product, or process is totally without risk. Consequently, at issue is whether those risks will be acceptable to society. Ultimately, the demand is for more information and understanding. If the situation continues and the risks remain uncertain, then the technologies and practices offered as part of agricultural biotechnology may never be acceptable to society.

But what are the risks and benefits, from environmental/biological perspectives, of agricultural biotechnology? What is known and not known about risk?

Environmental Risks and Benefits

Assessment of environmental risk is extremely difficult. It has become more difficult because of the tremendous impact that modern science has had (and will continue to have) on how we perceive biology. For example, the traditional proof of pathogenicity of an organism is based upon fulfillment of Koch's nineteenth-century postulates. These postulates state that the causal organism must be consistently associated with the symptoms of the disease, that it must be isolated and grown in pure culture, that upon inoculation the same symptoms must be produced, and finally, that the same organism must be reisolated from the inoculated host. All of these tests are qualitative, and although they were very useful when the development of the germ theory of disease was still underway, their value in many modern disease studies is often limited.

The change in concept from protecting an individual species to protecting a biological community will affect how the risks and benefits of agricultural biotechnology are evaluated.

Contemporary epidemiologists employ quantitative approaches to disease. These approaches have resulted in the acceptance of the tenet that long-term exposure to harmful agents, even at doses so low they cannot be detected, can increase the chances of disease. Advances in quantitative biology have also affected other relevant fields of biology.

Quantitative biology has had a significant impact on how we conduct and interpret biological experiments. The current paradigm posits that organisms exist within a continuum, rather than in a sharply delineated world where the answers may at first seem simple and straightforward. For example, as we realize the complexity involved in species survival, the concept of protecting a species has given way to the protection of biotic communities such as meadows and ponds. Consequently, the risk and benefit analysis has changed. These changes also affect how the risk and benefit of agricultural biotechnology are to be considered.

In general, the risks and benefits to the environment from implementation of biotechnology in agriculture can be grouped along a short-term, intermediate, and long-term time continuum. The short-term risks are usually restricted to those directly involved with the product at the site of manufacture or application. Their positive or negative impacts would be realized in the time period of hours to months. Intermediate impacts would affect closely associated organisms or functions over the following few years. Long-term impacts, occurring after five to ten years, are the most difficult to predict and could affect biological communities and our society at many different levels.

Short-Term Risks and Benefits

Short-term risks and benefits are defined as those that affect the specific production practice at or near the time of application. We can be most confident about our predictions when dealing with short-term impacts, for example, the use of an organism which produces toxic compounds during its growth.

When a toxin is produced by a living organism, toxicity studies are very difficult to conduct. Due to the ability of organisms to replicate and spread, they do not follow the classic approach to toxicology that has been developed for chemical substances which degrade over time and space. However, the ability of living organisms to grow and spread may also be the most beneficial attribute of the product. For example, long-term biological control of many insects and weeds has been achieved through a single or a few releases of predators, parasites, and/or pathogens of the pest.

Many of the inhibitory compounds produced by transgenic organisms can function as antigens. If a biological control agent or transgenic plant contains new or higher concentrations of antigens, it is possible that the increased levels of the antigen may enhance the possibility of allergenic responses. Determining the proportion of the population that would react unfavorably to these higher concentrations or new antigens is very complex.

Much of the early research in the development of transgenic plants has been to alter crop plants to make them resistant to pests. The success of such a product will be dependent upon its ability to produce compounds which are toxic to the pests. However, under certain conditions and at high concentrations, a risk is that the toxins could harm other life forms. From classical plant breeding programs we have already experienced varieties of crop plants which produced such high concentrations of alkaloids for the inhibition of insect pests that the crop posed a risk to human consumers.

Many apparently benign and commonly encountered microorganisms can often become pathogens under the proper conditions. For example, many bacteria and fungi known to be human pathogens also live as saprophytes in air conditioning and plumbing systems, on plants and plant debris, and in the soil. Also, many human pathogens are not identified until after repeated or high exposure occurs, as with the bacterium which causes Legionnaires Disease. This is especially true when other complications have inhibited the natural functions of the immune system. Many burn victims die from a bacterial infection caused by the same bacteria which are extremely common on fresh vegetables.

There are several possibilities available to manage the short-term risks that may be associated with biotechnology. A risk-management procedure would use methods to reduce the size of a hazard, such as appropriate protective clothing, handling, and application procedures. Formulations and delivery systems which minimize the chance of spills or drift are already well developed in the agrichemical industry.

With living products, the use of killed or sterilized organisms is possible. Genes which produce compounds toxic to pests can be inserted into bacteria and cultured under controlled conditions. The risk of spread and multiplication can be eliminated by killing the bacteria before application. Consequently, the more traditional risk assessment of toxic substances could apply. It also means that the product will have to be reapplied, ensuring a long-term market for the manufacturer.

Several products under development through biotechnology involve the development of herbicide-resistant crops. The risk associated with such products include the possibility that growers would actually become more dependent upon chemicals, since the resistance would allow the grower to apply the herbicide during the growing season, thereby reducing the need for conventional cultivation for control.

The most promising benefits are a decrease in the costs of crop production and the increased potential of growers to use minimum tillage methods. Minimum tillage operations, which are extremely effective in soil conservation practices, are often hindered in areas where weeds are severe. Although the use of herbicides may increase, soil erosion could decrease. Also, it has been shown that minimum tillage practices reduce chemical runoff, thus fertilizers and pesticides which are applied stay on the site of application rather than entering the surface water system.

The short-term benefits of the application of biotechnology to the environment may often be straightforward. The introduction of a plant variety resistant to disease or insect pests may reduce the need to use toxic chemicals, thus minimizing the negative impact on the environment as well as reducing production costs for the grower. Another short-term benefit of biotechnology products is to expand crop production areas by making it possible to grow salt-tolerant crops or crops not as susceptible to frost damage.

Intermediate Risks and Benefits

Intermediate risks and benefits are defined as those that affect the areas surrounding the site of application in a time frame of one to five years. In contrast to short-term risks and benefits, the intermediate impact of biotechnology on the environment is much more difficult to assess. Intermediate risks could include the eventual breakdown of the efficacy of a biological control agent due to selection of resistant populations, impact of residues from products of biotechnology on humans and the environment, and spread of unique genes beyond the site and species for which they were intended.

The ability of pests to adapt to agrichemicals and resistant crops is well documented. However, what is not known is how quickly the barriers provided by transgenic plants will be overcome. Many believe that the ability of pests to overcome products from biotechnology will be similar to our experiences with agrichemicals and resistant crops, where resistant pest populations are common.

The ability of the pest to adapt is dependent on the generation period of the pest, the genetic plasticity of the pest, the intensity of the selection pressure the pest is subjected to, the specificity of the mode of action, and the time (both duration and stage of the life cycle) that the pest is exposed to the control measure. The intensity of these factors will affect how quickly the pest will adapt and how successful the new population will be in causing economic damage.

The bacterial biological control agent *Bacillus thuringiensis* (often referred to as *Bt*) provides an example of intermediate impacts associated with biotechnology. The case of *Bt* illustrates the principle of transgenic plants which are able to produce the same toxin that provides the control. Preparations of the *Bt* bacteria have been produced and applied to crops for the last three decades to control harmful caterpillars. More recent strains include those that show promise in controlling mosquitos, Colorado potato beetles, and Japanese beetles. This is probably the best example of biological control of an insect pest based upon reapplication of the control agent.

Thus far the insect pests have not been able to overcome the efficacy of the *Bt* by developing resistance to the toxin produced by the bacterium. This is believed to be due to the general nature of the toxin, which degrades the gut lining of the caterpillar after ingestion, and to the short time period that the caterpillars are exposed to the toxin because the bacteria do not survive on plants.

The goal of many biotechnology efforts is to incorporate the bacterial gene that produces the toxin into plants or saprophytic bacteria that can survive on or in the plant. The risk of resistance

Use of minimum tillage systems that decrease soil erosion could be expanded with the advent of herbicide-resistant crops.

developing in the pest populations would be increased when these inventions are put to practice, since the pests would be exposed to the toxin for longer periods. In fact, in laboratory studies many of the caterpillars have already developed resistance to the toxin after exposure durations which simulate field conditions. The negative impact would not only affect the investment of the products, but would also result in the loss of efficacy of an extremely important biological control agent.

Methods to manage this intermediate risk have been proposed for the transgenic bacteria and plants. Transgenic bacteria can be applied after they are killed and still be effective. Subsequently, the toxin breaks down and is not continuously produced. With transgenic plants, it may be possible to engineer a genetic switch in which the plant would produce the toxin only at certain stages of its development and only in those plant parts which require protection from the insect. However, there would still remain the risk of consumers being exposed to high levels of potentially harmful compounds.

Residue analysis, as with toxicology analysis, is often based on the assumption that the residue does not increase over time and does not spread on its own. However, with produce from transgenic plants it is likely that the gene products will continue to be produced until the death of the plant cells. It is also possible for viable genetically engineered microorganisms to begin to grow whenever conditions become optimum, which also could occur long after harvest.

The matter is further complicated, for, as with current policy of affecting chemical residues, the total potential exposure must be considered. This includes not only the amount on a particular product but also how much of the product is consumed. Furthermore, it will depend upon which other products contain the residue and how much of each is consumed by an average individual. This requires knowledge of what crops are being treated, how the produce is handled, and the dietary preferences of the population.

Again, the very risks associated with a new technology are similar to the benefits. The potential to reduce the consumption of toxic compounds which we now rely on to produce our crops is a very important aspect of biotechnology. Through more efficient nutrient assimilation and partitioning, resistance to pests, and increased storage potential, the need for applications of pesticides would be reduced. However, the implementation of many products of biotechnology now under development will involve the release of genetically novel organisms into our environment.

Microorganisms such as bacteria can have a reproduction rate of once every 20 minutes. This phenomenal capacity to multiply is complemented by the ability of bacteria to transfer genes rapidly through populations via plasmids. Plasmids are segments of DNA separate from the bacterial chromosome that can be rapidly transferred to closely and distantly related bacteria. These properties make possible the extremely rapid buildup in hospitals of populations of *Staphylococcus* which are resistant to a number of antibiotics.

Another management of the risk of unwanted gene transfer is done by inserting the alien genes into the bacterial chromosome rather than into a plasmid. It is also possible to attach necessary promoters to other parts of the chromosome so that the gene cannot express itself without the presence and activation of the distant promoter genes. Separating the genes decreases the chances of both being inserted into another organism.

The risk of the transfer of altered DNA may also exist in plant and animal populations. However, the chance of gene transfer to other species is minimized due to the complexities of their sexual mechanisms. In temperate food-producing areas, the chances of gene transfer into native populations is even more reduced because so few of our food crops have closely related species in the same area of production.

In tropical areas, where many of the food crops are closely related to native plant species, there may be an increased risk of outcrossing and unwanted gene transfer. However, many of the same genetic techniques which would apply to bacterial populations, such as the requirement of complex promotor genes on distant chromosomes, may be applicable to plants and animals.

Long-Term Risks and Benefits

Long-term risks and benefits are defined as those that affect a large area and may not occur for several years after application of a product or practice. The assessment of long-term environmental risk is by far the most difficult and complex. How biological communities, such as a pond or meadow, and the abiotic environment, such as temperature, respond to human disturbance is extremely difficult to assess. For example, it is even more difficult to determine the impact of anthropogenic activities on the diverse microbial communities which are responsible not only for many animal and plant diseases but also for recycling nutrients in debris to make them available for future generations.

Long-term assessment is dependent upon our understanding of how natural and disturbed populations or communities function. Unfortunately, this is not well understood in most instances, especially when dealing with microorganisms.

A population, as defined in ecological terms, is a group of organisms of the same species occupying a similar space, allowing for the potential exchange of genetic information among the individuals. Some important properties of populations include density (number of individuals per area), natality and mortality (birth and death rates), age distribution, and dispersion. Genetic properties of populations include adaptiveness, reproductive fitness, and persistence. To adequately assess the potential impact of a product of biotechnology on a population, all of these factors should be considered. If released, it is also important to assess the population of the genetically altered organism. All of these properties of populations may be affected by characteristics of the population itself or the response of the population to its environment.

Population density is perhaps the most important property to consider, for it is the cumulative result of all of the other properties within a given area. However, assessment of density is not as straightforward as it may seem, especially when dealing with individuals that are difficult to detect due to clumped distribution patterns, low densities, small size, or rapid movement among sample sites. Even the densities of our well-studied deer and duck populations have margins of error.

The age distribution of a population is an indication of the general well-being of the population and a predictor of future growth trends. A population with few individuals of reproductive age would not be expected to increase quickly. However, if the population also had a large percentage of young, it would indicate a future potential for increase. If the population consisted predominately of postreproductive individuals, there would be little chance of rapid growth.

Whereas population density and age distribution would be important indicators for determining the impact of a genetically altered organism on the ecosystem, other properties are more relevant to determining the potential risks associated with the release of the novel organism.

Dispersal properties of a population will determine how quickly the population can spread to new locations or repopulate locations from which it had been removed. Control of dispersion restricts a population to its site of release, thus minimizing the potential of negative impacts on the surrounding ecosystem. Although restricting dispersal of individuals may be possible with physical barriers, it is more difficult to restrict the dispersal of pollen or seeds which contain novel genes.

Adaptiveness of a population is extremely important but is difficult to predict due to our limited understanding of how adaptive genetic traits function in natural systems. A population with high adaptive potential could

Assessment of long-term environmental risks and benefits of agricultural biotechnology depends upon our understanding of how natural and disturbed populations function.

be able to respond successfully to adverse conditions that previously would have been deleterious, such as the development of resistance of a pest population to a pesticide. It may be possible to restrict the adaptive characteristics associated with a specific gene by associating that gene with deleterious traits such as sterility.

Reproductive fitness and persistence are extremely difficult to predict and for the most part are passed on as observations after the fact. It will probably not be possible to predict these properties of a population without actual release followed by careful monitoring.

Biological communities can be defined as part of an ecosystem in which energy or matter is transferred among populations. Properties of communities include structural and functional aspects. The structure of a community includes species structure, as determined by the type of species, the diversity of species, and how stable the species composition is. Functional aspects of a community include the properties of energy and nutrient flow.

The community level of organization is very complex, and monitoring the effect of perturbations on the community often includes indirect as well as direct methods. For example, the displacement of a species may result in the invasion of several new species, which would affect the structural aspects of the community. If one of the invading species was a legume and able to fix soil nitrogen in association with *Rhizobium*, it would also affect the functional aspects of the community by affecting nutrient flow.

Several projects using biotechnology address the challenge of extending the host range for nitrogen-fixing bacteria common to legumes to include corn and other crop species. This would obviously have a positive impact by reducing our dependence on petroleum-based fertilizers, but would also have a significant impact on the nutrient flow in the soil on which the plants are grown. The possibility of nitrogen-fixing bacteria becoming associated with weeds or other non-target species must also be considered. To accurately assess the long-term impact of any project, one must consider the specific community and population characteristics that are important.

Designing Public Policy for Agricultural Biotechnology

The new techniques associated with biotechnology in agriculture are viewed by some as posing potential risks. These risks may be unknown, involuntary, uncontrollable, and capable of having extreme, irreversible consequences. If sufficient controversy is present, the general public perceives high risk, even if scientific and statistical evidence to the contrary is presented. While communities can readily accept "natural hazards" such as earthquakes or tornadoes, they are often less accepting of risks associated with new technologies.

But risks must be acceptable to us. Much of the controversy surrounding biotechnology is caused by the perception that the risk is not controllable by us. As a result, the public is increasingly insisting that government scrutinize and evaluate these risks before giving approval of the development and commercial marketing of biotechnology. Regulators are therefore forced to identify potential risks, assess potential impacts, and assure the public that the risks can be managed. Determining whether risks are acceptable is a two-stage process of risk assessment and risk management.

People can readily accept "natural hazards" such as tornadoes, but they are often less accepting of risks associated with new technology.

Despite federal and state efforts, however, a unified approach for regulating biotechnology has not been satisfactorily developed. As is often pointed out, a consistent and coordinated regulatory environment is necessary for the design of public policy that reduces the costs and uncertainties associated with the regulatory environment while assuring adequate protection for the environment and human health.

Public policy can be approached in two forms: regulation and judicial law.

Regulation is the imposition of constraints on the potential injurers by restricting their operations. This is known as an *ex ante,* or "before the fact," approach, and the imposition is *independent* of the occurrence of a specific event. It is intended to reduce the size and/or the probability of an event occurring. Judicial law is the tort law system of negligence, nuisance, liability, and trespass. This is known as an *ex post* or "after the fact" approach. It is *dependent* on the occurrence of a specific event, and it is intended to compensate the victim(s) or to recover damages for the harm done in the case of an event occurring,

These two approaches imply very different allocations of the risks created by hazardous events (Segerson, 1986). They therefore impose quite different conditions for firms investing in agricultural biotechnology and for the agencies responsible for regulating them (Larson and Knudson, 1990). Important questions are raised. What is an appropriate design of environmental policy? What are the costs of over-regulation? What is the potential impact of under-regulation? Is it possible to design policy that will both allocate risk efficiently and provide incentives for an optimal level of prevention or precautionary action at reasonable cost to the public and to the firms involved?

These same questions have arisen in debates over environmental policies towards hazardous wastes (Segerson, 1986; Johnson and Ulen, 1986; Cooter, 1986) and in considerations of public regulation of agricultural biotechnology field tests (Larson and Knudson, 1990). While the debate continues, the following general conclusions circumscribe the issues related to the environmental risks and benefits of agricultural biotechnology.

The socially efficient allocation of risk-related damages depends on who is more risk averse: the potential injurer or the potential victim. However, there does not appear to be any single approach to the allocation of risk damages that will achieve both an optimal allocation of risk damages and an optimal incentive for hazard abatement (Segerson, 1986). That is, it would appear that it requires both a judicial law and regulatory approach to achieve any sort of efficient public policy solution. This is a consistent conclusion, since it usually requires two policy instruments to achieve two policy goals.

Judicial Law Approach

Within the realm of judicial law, the efficiency of the chosen rules depends upon whether the precaution is unilateral (precaution by only one party) or bilateral (precaution by both parties). Strict liability rules are usually socially optimal when precaution is unilateral on the part of the potential injurers, whereas negligence is socially optimal when precaution is bilateral on the part of both potential injurers and potential victims. In the case of the occurrence of a hazardous event caused by a new technology, the efficient precaution is usually unilateral (on the part of the injurers), and therefore strict liability is the socially optimal form of precaution (Johnson and Ulen, 1986).

In addition, strict liability is simpler to try because the problems of proof are less severe and therefore the cost of litigation (which must be initially borne by the victims) is lower than under the negligence rules. Furthermore, strict liability provides a continuing incentive for accident-reducing innovation by the potential injurers, whereas the negligence rule gives little incentive to restrict potential injury because the injurers can escape liability by satisfying the legal standard (Cooter, 1986).

While, in general, strict liability is socially preferable to negligence, under a strict liability rule the plaintiff must prove that the defendant caused harm in order to recover damages. This situation creates uncertainty in the outcome of the policy and can therefore result in incentives for precaution that are either excessive or deficient relative to the social optimum. That is, in very low probability events, strict liability rules will result in too little precaution. In this case, additional *ex ante* regulation can provide the additional precaution necessary to ensure socially optimal outcomes.

On the other hand, over-regulation may lead to over-precaution, which may discourage investments in agricultural biotechnology research and development and result in less-than-socially-optimal decisions by the firm. Alternatively, there are dangers from relaxing the standards of proof that plaintiffs must meet in order to recover damages, and therefore additional *ex ante* regulation may be necessary to meet the social optimum (Johnson and Ulen, 1986).

Regulatory Approach

Can *ex ante* regulations fully replace *ex post* liability standards? As Segerson (1986) and Johnson and Ulen (1986) have shown, the answer is a categorical "no"! Regulation increases the likelihood of a victim being compensated. However, regulations alone fail to achieve an optimal allocation of risk damages while at the same time creating incentives for a firm to choose optimal or efficient levels of precaution.

Johnson and Ulen (1986) suggest that in the case of causal uncertainty (and independent of the firm's size and certainty of wealth), a firm that is subject to strict liability will generally be induced to take less precaution than is socially optimal. Thus, regulation that is slightly less than socially optimal could be used to induce a firm to increase its precaution to at least a social optimum. However, they suggest that because the ideal mix of standards is currently unknown, the current pattern of rules and regulations amounts to ad hockery.

Impact of Policy on Biotechnology Firms

From the firm's point of view, both *ex ante* regulation and *ex post* tort liability approaches will have an impact on the decisions of the firm to invest in agricultural biotechnology. Larson and Knudson (1990) examine four different regulatory approaches to limit public risks from field testing of agricultural biotechnology products. Their analysis shows that the four approaches are least effective at controlling public risks and maintaining incentives to conduct research when the research firm is small in size, returns to research are uncertain, and the potential damages exceed the value of the firm.

This probably explains why small, private research firms are often forced to merge with larger firms in order to continue in existence. Alternatively, public regulation could drive relatively small firms out of the industry, encourage small firms to engage in more risky research, or create strong incentives for large firms to divest certain hazardous activities to limit exposure to large liability payments (Larson and Knudson, 1990).

Larson and Knudson (1990) delve a little deeper into this problem from the point of view of the firm. They distinguish three cases under which the firm must make a decision as to the level of precaution chosen in allocation of resources to a risky activity. They show that the incentives for precaution fall as the size of the firm falls relative to the returns from the risky activity. However, in each case, the firm's wealth and the option of bankruptcy can complicate these results. Since a firm's ability to compensate is limited ultimately by its total wealth and since the total wealth is dependent on the success of the risky venture, then smaller or less wealthy firms may have more incentives to conduct riskier projects.

Ex post liability standards cannot always guarantee a social optimum. Depending on the probability of a hazardous event, a firm may choose a level of precaution that is either excessive or deficient relative to the social optimum. But a firm's ability to compensate for damages also determines the level of precaution it will choose, which in turn is determined by the size of the firm, the certainty of the returns to the risky project, and the magnitude and certainty of the potential damages. In each case, social optimums can only be guaranteed or the level of risk minimized by the additional presence of *ex ante* regulations.

Thus, it would appear that an appropriate environmental policy, one which allocates risk-bearing efficiently, provides incentives for an optimal level of prevention or precautionary action, and does so at reasonable cost to the public and the private firms involved, requires both *ex ante* regulation and *ex post* judicial law.

The Importance of Information

Information is considered a central limiting factor in the development of policies to assess and manage risks. If information is inadequate, then the policies are also inadequate. Lack of information introduces uncertainty into all decision-making processes and could lead to serious under-investment in agricultural biotechnology research and development.

For example, a public lacking sufficient information on the likely outcome of a new technology may be unwilling to share the responsibility for its implementation, and, therefore, it may not be implemented. If regulatory agencies do not have sufficient information on the likely outcome of a new technology, they are likely to drive the price of information to a level that will

Figure 1. *Cost and Value of Information.*

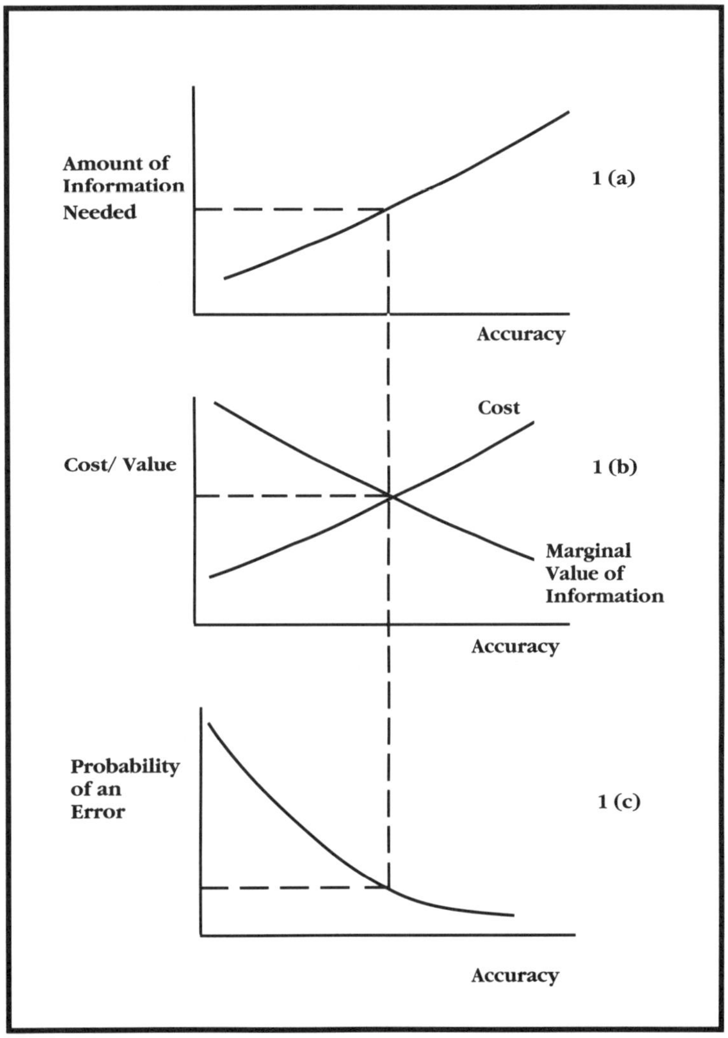

make it impossible to implement the technology. Finally, if the research firms do not have sufficient information on the likely outcome of a new technology or on the behavioral reactions of the public to a technology, then they may be either too risk averse to undertake the research or cavalier enough to ignore the consequences. In the latter case, bankruptcy may be the outcome.

Information has both a cost and a value. The relative tradeoffs between the cost and value of information are presented in Figure 1. Reducing the probability of an error requires increased accuracy (Figure 1c). But increased accuracy increases the amount of information that is required (Figure 1a) and also increases the cost of that information (Figure 1b). On the other hand, the marginal value of information decreases as accuracy is increased (Figure 1b). At some point, the extra cost of collecting or generating information exceeds the marginal value of the information and therefore limits its availability.

Apart from its cost, reliable information about the future is difficult to obtain. In addition, the further into the future the potential impact, the less reliable is the information, as in the case of long-term environmental risk.

There do appear to be ways of obtaining information that, in the absence of adequate information, could be used to evaluate programs that would regulate technology and information generation. For example, Conrad (1988) argues that contingent valuation methods (a benefit/cost technique) can be used to evaluate and design government policies, including regulations, that reduce risk.

However, the chief difficulty of applying contingent valuation techniques is the problem of obtaining reliable measures of willingness-to-pay for goods for which there are no markets. That is, answering questions like "What would you be willing to pay if this regulation were available?" is difficult for consumers when they have no experience with the regulations in question (Conrad, 1986).

A meaningful description of a government project on regulation and its future effects is very difficult because of the technical nature of the information required and the uncertainty associated with its outcome. For example, in a biological sense, it is not only difficult to quantify the interaction of populations in a steady state model, it is extremely difficult to assess when perturbations are introduced into the system.

Cooter (1986) raises two other factors that are important to effective public policy. First, there is a clear need to improve our understanding of the dynamic changes that occur as a result of alternative policies. Static equilibrium solutions do capture the crucial role of innovation, that it may often be responsible for environmental hazards but may also assist in reducing those hazards. Impeding technical change through high-cost regulations and/or litigation is an additional societal cost that is rarely accounted for. Or, as Larson and Knudson (1990) suggest, as firms use different technologies and strategies, the process of learning by doing combined with new experience and better information begins to improve the productivity of those technologies.

Second, Cooter introduces the notion of the distinction between prices (the cost of performing a permitted act) and sanctions (the cost of performing a forbidden act). He suggests that some environmentally hazardous activities must be priced (in the sense that they are a cost of doing business), and others must be sanctioned (in the sense that they are undesirable at any price). Regulators must decide whether the fine attached to violating a regulation is a price attached to the act or a sanction against wrongdoing.

Many of the solutions generated in these models ignore a number of factors that are important in determining the relative desirability of alternative policy approaches. Imperfect information and uncertainty notwithstanding, these include long latency periods, the role of insurance, uncompensable damages, and long-run responses by firms to alternative policies.

A Social-Psychological Perspective on Agricultural Biotechnology Risk

As assessment of risks moves further out on the time continuum, from short-term risks to intermediate risks to long-term risks due to accidental release, dispersal, etc., less is known. As long as adequate information is unavailable or inadequate, a situation of uncertainty exists and will likely persist.

Also, from social psychology it is known that the same information or reality can be interpreted differently, which may lead to uncertainty. Conflicting interpretations also lead to uncertainty. Much of the public attention and discourse on agricultural biotechnology focuses on environmental risks. The resulting debate between the "public" community and the "scientific" community is often confounded by different world views driven by distinct values, especially when dealing with unacceptable risk.

Unacceptable risk can be understood in two ways. From a risk management (not assessment) perspective, an unacceptable risk implies a probability of negative consequences in which the probability of loss must be reduced to an acceptable level. In contrast, an opposing view contends that any risk, no matter how small, is unacceptable.

For example, in the first case an agricultural biotechnology procedure that leads to a genetically altered organism, without adequate lab or field testing, may pose too great a risk to be released. Therefore, strategies and practices to manage the procedures for release must be developed to ensure that the unacceptable risk probability is reduced to an acceptable level (or even eliminated). In the second case, regardless of risk management strategies or practices, the release procedure presents an unacceptable risk.

The case of unacceptable risk seems to characterize much of the agricultural biotechnology debate. Different interpretations imply choices. As with most issues in the policy domain, choices are affected by perceptions, experiences, motivations, and values (or what is considered valuable). The dilemma illustrated by the above example provides entree to a social-psychological issue related to agricultural biotechnology: the debate over "real" and "perceived" risks.

It is truly unfortunate that the debate uses these terms. Such labels as "real" and "perceived" only serve to disguise the underlying nature of the controversy. Both types of labeled risks are real. Implicit in much of what is said about "perceived" risk is that such risk assessments are wrong, incorrect, or not real. These assessments are treated as perceptual distortions of what is real. The alternative interpretation is that "real" risks may be treated as distortions of fact or that the information used is unbelievable or inadequate. What separates them tends to be the value orientations.

Uncertainty due to interpretation of information is one of the core issues for the resolution of the environmental risks problem. The lack of reliable information in terms of environmental risks also presents challenges for scientists, public policy officials, and the public. Uncertainty due to lack of information also leads to problems for risk assessment and management. In both cases of uncertainty, the consequences may lead to oversimplification or distortion of the issues. In extreme cases of uncertainty, the result may be conflict.

Effective strategies for avoiding outrage must:

- *help earn trust, confidence, and credibility;*
- *enhance communication; and*
- *increase the accessibility of information.*

Conflict can lead to distrust and demands for absolute certainty. This may result in a demand for the no-risk option, as explained above. As defined by Sandman (1987), this is the concept of outrage. The outrage dimension is defined as everything about a risk that is relevant except how probable or likely it is to hurt you. Some factors that appear to underlie the outrage dimension include familiarity, control, dread, and moral relevancy.

The greater the outrage about any technology, including agricultural biotechnology, the greater the public concern about associated risks. If genetic engineering of microbes is viewed as an "outrage," the outcry against the risk of exposure of accidental release will be greater, even when scientifically little or no serious environmental hazard exists. If trust and confidence in scientists, regulators, and/or government are high, then outrage is low and the outcry against risk/hazard will be reduced.

If agricultural biotechnology is to result in economic and production benefits, then society's concern about environmental risks must be addressed. The public's confidence that environmental risks and production benefits have been adequately considered must thus be a concern of those responsible for agricultural biotechnology research, development, and implementation. From both policy and social-psychological perspectives, the call is to manage the controversies surrounding agricultural biotechnology such that outrage is avoided. The challenge is to avoid outrage, not to avoid the controversies.

What are some strategies for managing outrage? Effective strategies will include those which help earn trust, confidence, and credibility. Others will address methods for effective communication and strategies related to the availability and accessibility of information. All of these strategies must be continuously practiced. Although trust must be developed over time, it can be lost in response to a single incident.

The accessibility of information is basic to the goal of building trust as well as developing sound scientific data. Not only must the information be made readily available to interested parties, but it also must be presented within the appropriate context while leaving open the possibility of interpretation by the audience.

This is accomplished in part by presenting the relevant information in a clearly understandable manner and including the interested parties in the interpretation of the information and how it will affect the decision-making process. Information from private and public research and policy institutions must be made available to those interested, as well as the general public, to assure the development of effective policies which reduce risks and maximize benefits. These policies must be based upon trust and cooperation.

References and Additional Readings

Alexander, M. 1985. Spread of organisms with novel genotypes. In *Biotechnology and the environment: Risk and regulation. American Association for the Advancement of Science for the United States Environmental Protection Agency.*

Buss, D. M., H. C. Kenneth, and K.M. Dake. 1986. Contemporary worldviews and perception of the technological system. In *V.T. Covello and J. Mumpower (eds.), Risk evaluation and management. Plenum, New York, NY.*

Conrad, J. M. 1986. On the evaluation of government programs to reduce environmental risk. American Journal of Agricultural Economics 68:1272-1275.

Cooter, R. 1986. Liability rules and risk sharing in environmental and resource policy: Discussion. Amer. J. Agric. Econ. 68:1276-1278.

Davis, B. D. 1987. Bacterial domestication: Underlying assumptions. Science 235:1329-1335.

Douglas, M. and A. Wildavsky. 1982. Risk and culture: An essay on the selection of technological and environmental dangers. University of California Press, Berkeley, CA.

Ehrlich, I. and G. S. Becker. 1972. Market insurance, self insurance, and self protection. Journal of Political Economy 80:623-648. Freudenburg, W. R. 1988. Perceived risk, real risk: Social science and the art of probabilistic risk assessment. Science 42:44-49.

Freudenburg, W. R. 1988. Perceived risk, real risk: Social science and the art of probabilistic risk assessment. Science 42:44-49.

Johnson, B. B. and V. T. Covello (eds.). 1987. The social and cultural construction of risk: Essays on risk selection and perception. D. Reidel Publishing Company, Boston, MA.

Johnson, G. V. and T. S. Ulen. 1986. Designing public policy toward hazardous wastes: The role of administrative regulations and legal liability rules. Amer. J. Agric. Econ. 68:1266-1271.

Larson, B. A. and M. K. Knudson. 1990. Public regulation of agricultural biotechnology field tests: Economic implications of alternative approaches. Working Paper, USDA-ERS, p. 36.

Sandman, P. M. 1987. Apathy versus hysteria: Public perception of risk. In *L.R. Batra and W. Klassen (eds), Public perception of biotechnology. Agricultural Research Institute, Bethesda, MD.*

Segerson, K. 1986. Risk sharing in the design of environmental policy. Amer. J. Agric. Econ. 68:1261-1265.

Sharples, F. E. 1987. Regulation of products from biotechnology. Science 235:1329-1332.

Slovic, P. 1987. Perception of risk. Science 236: 280-285.

Wildavsky, A. 1988. Searching for safety. Transaction Publishers, New York, NY.

CHAPTER SIX

Many benefits are expected from agricultural biotechnology. To realize some of them, developmental phases, such as testing genetically modified organisms in the environment, will be necessary. The public wants assurance that hazards are eliminated or minimized. Taking some risk to obtain benefits is a necessary step in the sequence of agricultural research. However, biotechnology research has required special consideration because of the perception that it is particularly difficult to predict accurately how a genetically transformed organism will affect public health or the environment.

Currently, the federal government provides biosafety oversight and regulation through the Coordinated Framework for the Regulation of Biotechnology. Under the Coordinated Framework, certain federal agencies regulate the products of biotechnology, while others provide biosafety guidance and oversight. Some scientists say that this mixture of jurisdiction causes unnecessary confusion at the stage when it is appropriate to seek permission to conduct field tests with genetically modified organisms. This confusion and the regulatory burden on researchers appear to be the major causes of the slowdown in field testing by public-sector researchers and may ultimately result in the loss of the comparative advantage the U.S. currently holds in biotechnology research.

The public and the scientific community hold different views on the desired balance between the regulation of research and the freedom to investigate. How these differences will be resolved is a significant question that must be resolved if this nation is to allow agricultural science to move safely ahead with biotechnology applications.

6 U.S. Biosafety Regulations: Too Much or Not Enough?

D. R. MacKenzie, United Sates Department of Agriculture
Anne K. Vidaver, University of Nebraska-Lincoln

The Present Situation

The national investment in biotechnology research is substantial. The Office of Technology Assessment of the U.S. Congress placed the 1987 federal biotechnology research expenditures at $2.7 billion. Although relatively small when compared to the annual budgets of biotechnology research in human medicine, a more-than-modest portion of that federal investment is in agriculture. In Fiscal Year 1988 the U.S. Department of Agriculture spent over $100 million on biotechnology research. And this represents only a portion of the national total from all public and private expenditures in agricultural biotechnology research.

It is expected that the benefits of agricultural biotechnology research will flow to consumers as improved foods, fibers, and other products of commerce. This expectation has been given as the justification for the large amounts of public support expended annually in agricultural biotechnology research. Hopes have been expressed that, if done right, agricultural biotechnology will help increase U.S. farm profitability and address major concerns about the long-term sustainability of our national food and fiber production system.

Biotechnology also is expected to contribute to solving other problems related to agricultural production. Agriculturally related environmental problems such as nitrates in groundwater, pesticide pollution, and similar problems may be solved through biotechnology. Alternative technologies to heavy dependence on chemical fertilizers and pesticides are considered likely. These include the prospects for enhanced biological nitrogen fixation in crops through genetic engineering and improved crop and livestock protection through biological control practices. Biotechnology is also seen as critical for making U.S. agricultural products competitive in international markets. Through increased "value-added" technology as a result of bioprocessing, or through reduced production costs, biotechnology offers high hope for future U.S. agricultural profitability.

As with any new technology, concerns have arisen about its use. Some individuals and organizations in both the scientific and public communities are apprehensive about the safety of biotechnology research and the potential for genetically modified organisms to harm humans or the environment. These biosafety concerns are legitimate and represent a challenge both to the scientific community and to the public agencies charged with protecting the environment and public health.

Federal Regulation of Biotechnology

Contained Laboratory Research

Laboratory-contained research with a form of genetically modified organism derived by recombinant DNA is now pretty much a settled issue for safety. But, when recombinant DNA techniques (or genetic engineering) were in their infancy (ca. early 1970s), there was much conjecture about the uncertain hazards of the research. Some individuals called for heavy restrictions on the research. Others called for a complete stoppage. Some local governments (e.g., the city of Cambridge, Massachusetts) enacted legislation to regulate certain research activities.

State and national regulations were considered, but a solution was developed by the scientific community in an

As agricultural research in biotechnology has moved from the laboratory to the outdoors, a new series of biosafety questions have been raised.

historic meeting in Asilomar, California, in 1975. At that conference, a group of about 300 knowledgeable scientists proposed a plan that would operate through the National Institutes of Health (NIH). The plan called for a federally sanctioned scientific panel to judge the appropriateness of precautions used to conduct recombinant DNA research in contained laboratories. Different levels of containment were matched against the expected hazards of the experiment. From that proposal there arose the NIH Recombinant DNA Advisory Committee (NIH-RAC) and the NIH biosafety guidelines.

Over the years the NIH biosafety guidelines evolved to permit certain types of experiments to be reviewed by the NIH's delegated network of Institutional Biosafety Committees (IBCs). Other types of experiments are now exempt from review as they have come to meet the criteria for acceptable risk. More complex safety questions and uncertain hazards continue to be referred to the NIH-RAC for their scientific review and recommendation. This procedure has worked extremely well as a national biosafety umbrella for recombinant DNA experiments in contained laboratories. Through reasoned scientific judgment and experience, published guidelines are revised periodically. By working in open public meetings, the NIH biosafety review process has earned the respect of the scientific community and the trust of the public-at-large. Many of the earlier fears for the hazards of recombinant DNA research have been scientifically addressed, and approved experimentation has been allowed to proceed in contained laboratories.

Field Research

The problems of regulating experiments conducted outside laboratory containment are not settled, especially for field testing genetically modified organisms. In the normal sequence of agricultural research, laboratory discoveries advance from the laboratory to small-scale field research and eventually to commercialization. As biotechnology progresses, more and more research needs to be conducted out-of-doors. This has raised a new series of biosafety questions, for both the scientific community and the public-at-large, regarding adequate protection of the environment and the public health.

By the mid-1980s, biotechnology was again receiving close scrutiny to ensure that field research with genetically modified organisms would be conducted safely. A new system of biosafety review was needed to protect the scientific traditions of investigative freedom without unnecessary constraint or needless regulation while, at the same time, providing public accountability.

This need became particularly complex because of the National Environmental Policy Act (NEPA). Under the provisions of NEPA, any federal agency planning to take a major action that may significantly impact the environment must prepare an Environmental Assessment (EA). Originally intended to be a brief, easily readable document, case law and time have caused EAs to become highly complex technical documents requiring a great deal of effort to prepare.

Moreover, for those actions of a federal agency that will have, or will be likely to have, a major impact on the environment, an Environmental Impact Statement (EIS) must be prepared. These are long, highly detailed technical documents requiring major resources for development. Prior to the initiation of any proposed activity, the EA or EIS must be publicly announced (specifically in the *Federal Register*) and a period of time provided for public comment on the proposed action.

Inasmuch as a field research test might have a significant impact on the environment, NEPA has been held to be applicable for all federal funding of biotechnology research that is to be conducted in the field, as well as all field research regulated by federal statutory authority.

The Coordinated Framework

In 1984, the Domestic Policy Council of the White House announced in the *Federal Register* the "Coordinated Framework for the Regulation of Biotechnology" (49FR50856-50907). The final Coordinated Framework policy statements were published in the *Federal Register* in June 1986 (51FR 23302-23393).

The Coordinated Framework is based on the following three assertions.

- The *products* of biotechnology (not the *process*) would be regulated.
- Biotechnologically altered organisms do not differ fundamentally from nonmodified organisms.
- Existing laws would be used to regulate biotechnology; no new laws were needed.

Under the Coordinated Framework, several federal agencies asserted regulatory control over the products of biotechnology. For agricultural products and research, the predominant federal agencies have become the USDA's Marketing and Inspection Service (M&IS) and the U.S. Environmental Protection Agency (EPA).

The M&IS includes two agencies with regulatory authorities applied to biotechnology. The Animal and Plant Health Inspection Service (APHIS) is charged with providing protection from plant pests (Plant Pest Act) and regulating certain animal biologics (Virus-Serum-Toxin Act). The legal authorities of APHIS have been extended to cover the products of biotechnology through the existing application/permit process.

The Plant Pest Act currently is the basis for regulating genetically engineered plants because the commonly used vector systems and the genetic promoters used in plant biotechnology are almost exclusively from plant pests. However, none of these genetic derivatives lead to a plant pest or disease problem. The importation, interstate transport, or release into the environment of plant pests or organisms thought to be plant pests, or genetic sequences therefrom, are regulated and thus require an APHIS permit for each experiment or set of experiments. It is not yet clear how these regulations will apply to the new genetic transfer methods such as microprojectiles, electroporation, and similar techniques which do not use plant pests or, *sensu* NIH, even recombinant DNA methodology.

Under the Virus-Serum-Toxin Act, APHIS regulates products of biotechnology such as vaccines and other biologics produced through the new methods of genetic modification.

The second USDA agency that regulates biotechnology is the Food Safety and Inspection Service. This agency ensures the safety of biotechnology-derived livestock and poultry under its stringent standards, if they are to be used as human food.

The EPA has announced plans to regulate genetically modified microorganisms under two statutes. The first is the Federal Insecticide, Fungicide and Rodenticide Act (FIFRA), which regulates, among other things, the safety of tests with pesticides. Previously, research with chemical pesticides (i.e., nonliving) in small-scale tests (less than 10 acres of land or less than one acre of water) was exempt from regulation, but this presumed exemption may not be applicable to microbial pesticides. Experiments with microbial pesticides now require an EPA permit prior to initiation of the test.

The second EPA authority that may be applied to biotechnology is the Toxic Chemical Substances Act (TCSA). By considering DNA as a chemical substance, EPA proposes to extend its authority to cover genetically modified microbes that might be used in industrial applications such as bioremediation, bioprocessing, mining, and other uses. However EPA, under TSCA, defers to APHIS for most agricultural applications.

As of this writing, revised EPA regulations are undergoing preparation for public comment and subsequent promulgation. The agency is, however, issuing permits based on its original plan, published in the June 26, 1986, *Federal Register*.

Table 1 presents an overview of the status of federally approved field test applications as of January 1, 1990. Most of the permits have been issued to private-sector applicants, and the majority of approvals have been for tests with genetically modified plants.

Other federal agencies also assert regulatory authority that may affect agricultural research. The Food and Drug Administration (FDA) regulates food additives, drugs, and similar substances that may have agricultural application. The Occupational Safety and Health Administration (OSHA) oversees employee safety in the workplace, which applies as well to biotechnology research. The U.S. Postal Service regulates the shipment of some products, including biologics, that would be used in agricultural biotechnology research.

Jurisdictions and Implementation

The Coordinated Framework for the Regulation of Biotechnology has helped to establish jurisdictional authorities for biotechnology. To further assist in the clarification of jurisdiction, the Office of Science and Technology Policy established the Biotechnology Science Coordinating Committee (BSCC) (50FR47174-47195, November 14, 1985). Agency heads, as members of BSCC, meet regularly to address items such as jurisdictional authority and to assure that the Coordinated Framework is operating smoothly.

The Coordinated Framework is not without difficulties. Some areas of biotechnology research are not now covered by regulatory authority. For example, nonfederally funded research with transgenic fish in outdoor ponds is said to be not regulated by anyone.

Table 1. *Federally Approved Biotechnology Agricultural Research Field Test Applications (through January 1990).*

Research sector	Federal agency: USDA New	USDA Repeats	EPA New	EPA Repeats	Total
Private	23[1]	19	08[3]	10	60
Public	04[2]	01	02[4]	00	07
Subtotal	27	20	10	10	
Total	47[5]		20		
				Grand Total	67

Notes:

[1] *9 Tomato, 7 Tobacco, 3 Cotton, 2 Alfalfa, 1 Potato, 1 Soybean*

[2] *2 Tobacco, 1 Cucumber, 1 Poplar*

[3] *4* Pseudomonas spp., *2* Clavibacter spp., *1* Rhizobium sp., *1* Bradyrhizobium sp.

[4] *1* Pseudomonas sp., *1 Insect virus.*

[5] *102 total by January, 1991.*

Such examples have caused some state governments to initiate biotechnology legislation. This may prove to be undesirable inasmuch as it may create a patchwork of state regulations that will hamper biotechnology research and commercialization.

The Coordinated Framework has also led to some confusion for principal investigators due to perceived overlap of jurisdictions among federal agencies. The regulation of nonpathogenic organisms (such as *Rhizobium spp.*) under the Plant Pest Act is an example. The field testing of a genetically modified *Rhizobium* requires an EPA permit. Normally, APHIS defers to EPA on a *Rhizobium* request for review. But, if the genes from the *Rhizobium* were to be successfully introduced into a plant species, that test might also require an APHIS permit. Examples of confusion abound and have no doubt contributed to the present frustration expressed within the scientific community.

Another source of frustration for principal investigators has been in the slow implementation of the federal regulatory and guidelines structures. Several factors have slowed the process. Legal challenges and the threat of legal challenges have urged caution on federal agencies to avoid lengthy and costly litigation to sort out procedural questions.

Alternatives to Regulation

Field Research Guidelines

The federal regulations that apply to biotechnology field research are quite different from the biosafety guideline/ review system initiated by NIH in the early years of biotechnology. Regulations carry legal penalties, including substantial fines. Regulations are characteristically highly specific, and commonly, they closely define allowable activities and procedures, especially for products to be tested and sold in the marketplace.

Guidelines, on the other hand, are more easily changed as more information is acquired. And they do not carry the penalties of regulations. However, failure to comply with guidelines can result in the loss of research funds to the investigator and the employing institution, or their being sued for damages. Furthermore, guidelines have historically been the operational procedures for conducting research, especially by academic and non-profit institutions. Such research may never result in a product, but may be essential for scientific understanding of biological processes.

In recognition of this distinction, the USDA proposed to establish a voluntary system of guidelines for research outside of the laboratory. The rationale for this proposal included:

- The immense opportunities to apply biotechnology to agricultural problems;
- Recognition of field research as an important link in moving discoveries from the laboratory to eventual agricultural commercialization;
- The existing, extensive experience and knowledge available from conventional agricultural field research that is applicable to safe field testing with genetically modified organisms;
- The enormous demands that could be placed on the NIH system if their mechanism of oversight was to accommodate all of the national biosafety review needs of agricultural research; and
- The need for oversight of biotechnology research beyond recombinant DNA..

In 1986, the USDA established the Office of Agricultural Biotechnology (OAB) and, subsequently, the Agricultural Biotechnology Research Advisory Committee (ABRAC) to advise the Department on the safety of tests to be conducted outside of the laboratory with genetically modified organisms. The OAB supports the activities of the ABRAC. The ABRAC membership is made up of scientists and others knowledgeable on the issues of biosafety and the environment and is closely modeled after the highly successful NIH-RAC process.

The USDA system of guidelines is intended to serve as a complement to the NIH-RAC. The ABRAC is presently drafting guidelines that will then be circulated for public comment. It has also assisted the USDA-OAB in the development of a manual intended to help principal investigators understand the Coordinated Framework and the USDA's biosafety review program (Purchase and MacKenzie, 1990). Individual proposals have already been reviewed on a case-by-case basis, helping ABRAC establish the applicability of the draft guidelines.

Out of necessity, the ABRAC guidelines have approached biosafety questions for field research differently than did the NIH-RAC for laboratory-contained research. The RAC review process was based on the previously existing Centers for Disease Control classification of vertebrate pathogens. This is understandable inasmuch as the NIH wanted to provide assurance that proper precautions for laboratory containment would be followed, primarily to protect human health. The likelihood of an organism's escape from a contained laboratory was considered to be negligible, and hence there was less concern for environmental consequences.

The considerations assigned to ABRAC are clearly quite different. The opportunities for the escape, establishment, and consequent environmental impact of genetically modified organisms in field tests had to be assessed. The conceptual approach used by the ABRAC guidelines resulted in a five-step process.

The first step is the assessment of the hazards of the unmodified organism. In many cases, agricultural organisms are very well understood, both in their behavior and in their consequent impact on the environment. The greater the familiarity with an organism, the greater the assurances that can be given as to proper risk-assessment classification. This distinction differentiates between a known, high-hazard weedy species, like kudzu, and maize, which has virtually no capacity to become a weed in the U.S.

The second step in the guideline's assessment process is to make a judgment about how the genetic modification affects the original organism's potential for adverse effects. Some modifications may lower an organism's potential for hazard, while others may increase it. Some modifications may have no effect on the organism. This judgment is based upon prior knowledge of the incorporated genetic material, preliminary tests of the genetically modified organism in a contained environment, and on the best scientific judgment available to the investigator.

The third step is to assign a safety "condition" or changed status to the genetically modified organism, based on the assessments made in the first two steps.

The fourth step in the assessment process is to make reasonable proposals for the types of confinement that will be used in the field tests. Organisms can be confined in field research by using physical, chemical, biological, environmental, and scalar practices that greatly reduce the likelihood that an organism will cause adverse effects on the environment or public health.

The fifth and final step determines the appropriate level of scientific review for the proposed biosafety protocol. Potentially hazardous tests, or those with a large degree of uncertainty as to the organism's effects, receive ABRAC review. The less hazardous tests are reviewed by the local Institutional Biosafety Committee. Highly hazardous tests would not be allowed outside laboratory containment.

An evaluation of the safety of the planned research, along with a description of the test site, emergency procedures, plans to terminate the experiment, and other critically important information, is used by ABRAC to make a recommendation to the USDA's Assistant Secretary for Science and Education.

Scientific Support

Another activity in the USDA has been the establishment by the Cooperative State Research Service (CSRS) of the National Biological Impact Assessment Program (NBIAP) to facilitate the safe field testing of genetically modified organisms. The NBIAP was first proposed by the National Association of State Universities and Land Grant Colleges' Committee on Biotechnology to address research needs of the public and private sector research communities.

The underlying assumption of the NBIAP is the well-recognized fact that there exists a tremendous body of knowledge on the methods and procedures of field testing from the conventional research of the past. This knowledge is applicable in many ways to the questions of safe field tests for genetically modified organisms. It is the purpose of NBIAP to facilitate the appropriate application of this body of knowledge to biotechnology and to develop new methods to assure safe and productive field tests with genetically modified organisms.

To accomplish its objectives, NBIAP developed a strategic plan which identifies three areas of activity. Those areas are:

- Information network
- Biological monitoring
- Biosafety research

Information Network

In collaboration with a number of institutions, NBIAP has developed an information network to support biosafety needs in both the public and private sector research communities.

The information network is made up of an electronic bulletin board that gateways into a family of data bases on biosafety topics.

Electronic Bulletin Board

The information network participants link via their computer terminals and modems to the NBIAP electronic bulletin board, which provides up-to-date information on biosafety-related research activity and announcements. Provision has been made for broad participation on the bulletin board and for individuals to communicate with each other via an electronic mail (E-mail) service. Thus, individuals can use the network to contact other scientists working in similar areas to obtain assistance when planning field research with genetically modified organisms.

Data Bases

The electronic bulletin board also serves as the gateway to 14 data bases that were designed to provide accurate and up-to-date information on a variety of topics. The purpose of the data bases is to assist principal investigators in securing authoritative and scientifically valid information. Some of the data bases are described below.

The NBIAP is directly supporting several of the data bases. One of these data bases contains the current text of all federal laws, regulations, rules,

Scientists can use NBIAP's electronic bulletin board to get the latest biosafety-related announcements, to contact other scientists working in similar areas for assistance when planning field research with genetically modified organisms, and to access data bases containing a broad range of biosafety information.

guidelines, and Executive Orders pertaining to biotechnology and biosafety. The system also contains a current data base of all public and private Institutional Biosafety Committees (IBC), including the names of contacts, addresses, telephone numbers, and facsimile numbers. NBIAP also maintains a current data base on all approved applications to federal agencies related to field-test permits, licenses, and scientific reviews. Proprietary or confidential business information is not included.

The USDA's National Agricultural Library (NAL) is supporting three of NBIAP's information network data bases. The first one is the "Yellow Pages," which is a directory to other directories. The types of information included are bibliographic and nonbibliographic listings, such as videotapes on field studies, and other data bases. A second NAL-supported data base contains current biosafety literature. This bibliographic file provides information on currently available books, reports, and newsletters. The third NAL-supported file lists recent U.S. patents that have been awarded for genetically engineered species.

Other data bases are maintained in cooperation with the North Carolina Biotechnology Center and with network participants. These include information on topics such as new jobs and positions available, biotechnology companies, state information, a directory of plant biotechnology scientists, and more.

Collaborators at the University of Arizona recently completed a national survey of public-sector research experiment stations that compiles edaphic, climatic, geographic, and commodity information for each site. This information has been added to the NBIAP data base system to further assist scientists in selecting sites for safe experiments with genetically modified organisms.

A proposed data base will cover the organisms of biotechnology for the purpose of systematically accumulating the experience and knowledge of previous field tests. This will be used to assist scientists in planning safe experiments, serve as a resource for regulatory agencies to help them make scientifically sound decisions, and provide the public with a source of documented evidence on the outcome of experiments conducted outside of contained laboratories.

Knowledge Base

A third component of the information network is the knowledge base, which is designed to assist principal investigators in preparing applications for permits, licenses, and/or scientific reviews. The knowledge base contains two expert systems. The first identifies the responsible federal agencies under the Coordinated Framework to which the application(s) should be directed. The second expert system assists principal investigators in evaluating the levels of laboratory containment required under the NIH guidelines. Other expert systems are planned and will be undertaken as resources permit.

The knowledge base also has an educational component of information drawn from broad sources, but it is focused on issues in biosafety and field tests. This text has been organized into a "hypertext." Hypertext is a computer-based, hierarchical presentation of information ranging from general to specific explanations that can be interactively selected by the user. Through hypertext, the reader can pursue information to his or her desired level of detail.

The third segment of the knowledge base is the "intelligent form generator." This unit of the knowledge base is used to develop first-draft applications for federal permits, licenses, and scientific reviews. The intelligent form generator presents to the user menus of experimental design options for selection, as appropriate to the proposed test. This information has been assembled from existing knowledge (expert panels, literature, and previously approved applications). Users have access to previously written standardized text, technically precise descriptions, test-site information, and other data base resources.

By combining data base information with extensive menus specific to the organism categories that have been designated for agriculture (e.g., pond contained fish, cotton, nonpathogenic soilborne bacteria), specific research methods, procedures, and protocols can be brought together to form a technically precise application. Eight organism groups are included in the first version. Eventually this will be 79 groups to cover all of agriculture. The software application offers a word-processing option that allows re-drafting of the materials for clarity, as well as scientific and technical accuracy.

The purpose of the intelligent form generator is to assist the principal investigator by lessening the burden of complying with federal biosafety oversight requirements while at the same time not compromising those requirements. It is to be hoped that, with this intelligent form generator, safe field testing will be facilitated through assisting scientists in the exchange of biosafety information and through helping them in the application for federal permission to conduct field tests.

Biological Monitoring

Another activity of the National Biological Impact Assessment Program is the facilitation of biological monitoring for genetically modified organisms tested outside of contained laboratories. To do this, NBIAP has undertaken a national survey of previously conducted field studies. The survey will gather the existing scientific experiences and then organize that information for the benefit of other scientists and science in general. This research-based reporting system will undoubtedly become central to future efforts to place the hazards of biotechnology research in balance with appropriate levels of oversight regulation. As a resource to the scientific community, the information on what is known and what is available is seen as a valuable service, one that is certainly consistent with the mission of NBIAP—to facilitate safe field testing with genetically modified plants, animals, and microbes.

Biosafety Research

The National Biological Impact Assessment Program is also providing modest financial support for specific biosafety research projects. Biosafety is seen as a research topic unto itself. Although much useful information can be derived from conventional research, some aspects of biotechnology research raise unique biosafety questions. New biosafety methods must be developed for those needs. This will require new predictive models to understand the likely consequences of a field test with a genetically modified organism. This information will be critical to establishing the expectations for commercial releases of a biotechnology product, prior to that activity.

Improved biosafety methods and accurate prediction models may represent the best current approach to providing biosafety assurance in the long term. This will, however, require a more thorough understanding of how organisms interact and how varying environmental conditions affect movement and dispersal patterns; how physical factors affect survival, fitness, and reproductive behavior; and how all these come together to impact the environment or public health. More knowledge on these topics should lead to better biosafety assurances and more expeditious applications of biotechnology to agriculture.

Cost of Assuring Biosafety

The biological hazards of biotechnology remain conjectural; this point requires emphasis. However, incorrectly predicting the biological hazards of biotechnology can have serious economic and social consequences. For example, a genetically modified organism wrongly thought to be safe could have major undesirable consequences and inflict considerable economic damage on the environment or public health. Such "worst case scenarios" often draw upon the absence of good scientific knowledge or are based on unscientific assertions about recombinant DNA or other processes. This has contributed to some confusion in the popular and even some scientific literature regarding the true risks of biotechnology research.

But incorrect predictions can also result in overly cautious restriction and, hence, unnecessarily limit biotechnology research. In such a case, valuable research would not be undertaken because of unjustified or unrealistic fear. It is difficult to make a current assessment of these two extremes because of a lack of good scientific knowledge on the biosafety issues of biotechnology. It is generally recognized that some research will be unsafe and should not be undertaken. It is also generally recognized that some experiments will be acceptably safe. To not undertake these acceptably safe experiments, or to do so at a slowed pace, will result in costs to our national economy as lost or delayed opportunities.

The regulation of biotechnology is not scale-neutral. There are transactional costs in the application-for-a-permit process. Private-sector research is alleged to be more accepting of the regulatory process because professional assistance is provided when preparing applications for permits and the costs are passed onto others. Small companies are said to carry a greater relative burden in preparing applications than do large companies.

Public-sector research places the burden for any regulatory applications squarely on the shoulders of the principal investigator. Federal funding for research projects requires additional biosafety reviews that may impose even more burdens on some scientists.

All of these considerations have the effect of increasing the economic consequences of regulation, without necessarily assuring biosafety, and decreasing the willingness and responsiveness of the scientific community, especially at academic institutions, to undertake certain types of field research. Federal agencies, as well as the public, have difficulty dealing with relative measures and probabilities that form a necessary part of scientific research.

The regulation of biotechnology is not scale-neutral.

National Survey

A recent national survey investigated the impact of biosafety regulation on research with genetically modified organisms (Ratner, 1990). The survey covered biotechnology research areas such as human health, environmental protection, and agricultural and industrial application. Two-thirds of the respondents were employed in the public sector, and the distribution of research activities covered plants, animals, and microbes. The respondents were predominantly funded from federal sources (approximately two-thirds).

Seven percent of the respondents had already conducted field tests with genetically modified organisms. Permits were obtained from NIH (two), USDA (13), EPA (11), and state authorities (7). The total reflects some redundancy in permit requirements and one test that did not require a permit.

Of the 345 respondents to a question on whether or not they had a genetically modified organism sufficiently advanced for field trials but elected not to proceed with the test, 12 percent responded "yes." The most common reasons given as important or very important for the decision included legal constraints (67%), regulatory uncertainty (81%), too much paperwork required (76%), and too much time required (71%). The protection of patent rights did not seem to be an important concern.

A total of 355 responses were given to the question "Have you been discouraged from conducting field trials with genetically modified organisms?" Sixteen percent of the respondents indicated they had been discouraged. The sources of this discouragement were clearly press reports and regulatory uncertainty. There was no indication of significant discouragement from colleagues, institutions, or professional societies.

When asked if they planned to conduct a test outside a contained facility with a genetically modified organism in the next two years, 25 percent indicated that they planned to conduct such tests. Responding to a question on whether or not they would be required to obtain a permit, license, or some type of approval to conduct the field test, three-quarters of the respondents said "yes." When asked "From whom?" one-third indicated APHIS, one-quarter indicated the U.S. EPA, and the remainder indicated other agencies, including the Food and Drug Administration, local, foreign and unknown.

The results of this survey clearly indicate a more than modest, and probably growing, regulatory burden on the U.S. biotechnology research community for field testing.

The Regulatory Burden

If the U.S. regulatory system has come to be seen as overly burdensome, there is a fear that some researchers will seek alternatives (such as foreign collaboration) to escape the U.S. regulation of biotechnology. Alleged examples of this have been reported in the popular press. An economic consequence of foreign collaboration to avoid regulation is the likely loss to the U.S. of the commercial advantage to be derived from domestic research. Invitations from foreign governments to U.S. scientists to conduct research abroad without the burden of regulation is also a concern for U.S. policy makers.

Multinational corporations offer a similar dilemma. Multinationals may pursue opportunities that offer the least regulation and maximum efficiency. If the burden of U.S. regulation on biotechnology research is perceived by multinationals to be too great, significant portions of their experimentation may be driven elsewhere. This, too, could have negative economic consequences for the commercialization of biotechnology in the U.S.

Efforts are under way with our major trading partners to coordinate protocols for field research. The European-based Organization of Economic Cooperation and Development is working closely with the U.S. government to develop good research practices for field testing. Additionally, the European Economic Community is working with the office of the U.S. Federal Trade Representative to try to coordinate protocols for biotechnology field testing in an attempt to avoid disharmony in the international requirements for product registration, such as the problems that have evolved for drug and pesticide marketing.

A potential negative effect of these agreements is the concern for intrusion into the traditional freedom of the research process that allows the investigator to select the most appropriate procedures for individual experiment, rather than operate by prescribed protocols. Standardization also has economic consequences that must balance the uniformity of international research protocol agreements against the experimental design flexibility required or desired by individual investigators.

Biotechnology brings new choices and challenges to research and commercial product development. The points described above, plus the need for protection of intellectual property, the necessary exchange of scientific information, and many other issues, must be addressed. This might be done through international treaties designed to foster and protect not only the research process, but the capital investment made in biotechnology. International treaties could facilitate the international exchange of information on biosafety, better coordinate biotechnology projects (e.g., genomic mapping), and provide improved patent protection for novel ideas and products that have been genetically modified.

Societal Perceptions of Regulations

Different sectors of society have different perceptions of the protection afforded by federal biotechnology regulation. In general, it is asserted that consumers tend to accept the regulatory process as reasonable assurance of safety.[1] The regulation of drugs, the inspection of food, the registration and inspection of motor vehicles, and similar types of mandatory controls have come to be accepted as part of everyday life in America. Those freedoms protected by the U.S. Constitution are jealously guarded as civil liberties. Not much thought has been given by the public-at-large to the regulation of research. There are some public concerns about problems that are perceived to result from new technology. Hence, the public calls for greater oversight and regulation of research.

In general, the research community has felt that it has historically policed itself in a responsible way. For example, agricultural field research has been virtually without a major safety incident for more than 100 years. Both in the public and private sectors, researchers have long cherished the freedom to investigate problems without the heavy burden of regulation. There have been some relatively new requirements that have been well accepted by researchers, such as radiation safety, human subjects protection, and similar seemingly necessary intrusions into the research process.

Heretofore, a "presumed regulatory exemption for science" has been extended to the research community in areas such as pesticide testing, varietal release, conventional genetic experi-

[1] *However, this attitude may be changing, as indicated by the success of some supermarkets that are retesting produce using standards higher than those of the federal authorities and some chains that are declaring their intention not to sell dairy products from cows injected with bovine somatotropin.*

ments, and similar investigations performed on a limited scale. Even though some of this research is conceivably hazardous, the judgment of the principal investigator and the responsibility of the sponsoring institution were accepted as sufficient protection. This sort of research activity has been accomplished without incident, and the results have been made available to the public and the scientific community.

Private-sector researchers have become more accustomed to dealing with regulation than have their counterparts in the public sector. To them, regulation is considered simply as one of the costs of doing business. The registration of compounds, the compliance with strict regulations for commercial products, and similar regulatory needs have been accompanied by corporate acceptance of a large, complex federal regulatory structure. As noted earlier, corporate research units typically have, in addition to research scientists, a professional staff experienced in developing applications for permits or regulatory review.

The normal sequence leading from the laboratory to field research is now seen as a weak link in agricultural biotechnology. The new regulatory structure and the perceived and actual burden of making application for permission to conduct field tests have become an obstacle cleaving laboratory research from applied field investigations. This is a major concern for public institutions (such as universities) that do not have a professional staff to deal with regulatory requirements that are basically designed for product development. Some public institutions may respond by hiring one or a few professional staff writers. Other institutions may simply forego field research and do only greenhouse and growth chamber studies. This would have the effect of significantly slowing the commercial development of agricultural biotechnology.

Some nonprofit, public-policy institutions favor more stringent regulations, arguing that the present regulatory structure is incomplete and/or flawed. Previous experience with environmental disasters such as the Exxon-Alaskan oil spill, Three Mile Island, and other "regulated" technologies has done little to arrest their concerns. Some of the most outspoken public-policy institutions have called for a halt to biotechnology research until more information can be gathered regarding the safety of field experimentation with genetically modified organisms.

The complexity of attitudes towards regulations results in some curious reactions. Criticism by public-policy institutions erodes the public's confidence in the biosafety regulatory structure. The near-passive acceptance by the private-sector scientists of regulatory requirements is troublesome to the public-sector scientist. Public-sector scientists' attitudes towards regulatory requirements are seen by some in public-policy institutions as added justification for their distrust of *all* scientists. Contributing further to the public concerns for biotechnology regulation are the gaps in the Coordinated Framework and the logical inconsistencies as to who regulates what and even what is to be regulated (e.g., modified organisms derived by recombinant DNA or by other techniques, as well). Is a new national biosafety structure needed?

Because of the perceived and actual burden of regulatory requirements, the normal sequence leading from the laboratory to field research is now seen as a weak link in agricultural biotechnology.

The Future: Trends and Projections

A major policy issue that needs to be decided for the future success of agricultural biotechnology is whether biosafety assurances for agricultural biotechnology research, as opposed to product development, will be provided predominantly by regulation or by guidelines. If regulations are to be the method for providing biosafety, will the mechanisms and criteria be used for research as well as the products?

The guidelines process delegates responsibility according to an assessment of the safety and likely consequences of research with genetically modified organisms. Some types of research would be overseen at the local level by the Institutional Biosafety Committee (IBC), or some aspects even considered as exemptions.

No such mechanism has been developed for the regulatory processes, except for case-by-case petition. That petition process is seen as slow, cumbersome, and one not likely to be used by individual scientists or the public sector. Should a federal agency be assigned the responsibility of assessing biosafety and providing appropriate deregulations and exemptions? And, if so, who should do it?

A second issue that needs to be resolved is the appropriate amount of regulatory burden needed for the principal investigator. If the private sector has been able to cope with regulatory requirements through adding additional professional staff, should the public sector be provided the resources to do so? And, if so, by whom? Would additional resources only cover the direct consequence of the regulatory burden and yet leave in place many of the undesirable, indirect consequences of regulations (such as research areas that are avoided by scientists as too regulated) and/or contribute to research inefficiency? Who should assess these factors? Who pays the cost?

If a public institution's solution to the regulatory burden is to form private-sector partnerships, what will be the agreement on the intellectual property developed from the research? The benefits of having a private-sector partner to help cope with regulatory requirements could quickly be offset through the loss to the institution of patent rights and royalties. Will only apparently commercially useful research be done?

There are also significant implications for legal liability as a consequence of a biosafety error, should it occur as a result of biotechnology research. Policy makers need to address the implications of noncompliance with regulations versus institutional liability through negligence. Does compliance with the regulatory process and adherence to federal guidelines further reduce an institution's liability or only double the bureaucratic burden?

If the U.S. biosafety assurance process develops as an overly burdensome impediment to research, biotechnology will likely be constrained, and commercial development will be delayed or be unrealized in some areas. Or worse, some scientists may choose to defy the regulatory requirement and create a public relations spectacle that will injure the image of not only his or her institution, but the reputation of science as a whole.

If the federal regulatory structure develops some flexibility, research progress may be expedited and conducted with public support. How this might happen will depend on policies yet to be established and promoted by national leadership in both the public and private sectors. These policies should accommodate 1) the broad traditional freedoms of agricultural scientific research that have given this nation the world's safest food and fiber at the lowest cost to the public, 2) the needs and consequences of the new technology, and 3) the national and international implications of the effects and benefits of biotechnology research. To leave these issues and policies unaddressed and unattended will cause significant scientific and economic consequences for future generations.

Conclusions

Agricultural biotechnology research offers tremendous opportunities to benefit agriculture on many fronts. Concomitant with these opportunities are the responsibilities to provide adequate biosafety protection for the environment and public health. The Coordinated Framework has served as a reasonable interim plan. Regulation of the scientific process under the guise of regulating the "product" has become an unfortunate reality. Federal guidelines added to federal regulations are seen by some as "double jeopardy" and by others as the only scientifically defensible way to do research (i.e., not product development). A modified structure or new thinking is needed, along with possibly new legislation.

The products of biotechnology may not be fundamentally different from conventional products, but the present biosafety treatment of biotechnology research certainly is. The current and growing regulatory burden on science must become commensurate with appropriate safety assurances if this nation is to enable agricultural science to move forward. In the words of Thomas Jefferson, "Freedom is the firstborn daughter of science." In today's regulatory climate, one wonders if Jefferson didn't get it reversed.

References and Additional Readings

Batra, L. R. and W. Klassen. 1987. Public perceptions of biotechnology. Agricultural Research Institute, Bethesda, MD.

Brown, Bart, ed. 1989. New developments in biotechnology 5: Patenting life. Office of Technology Assessment.

Carlson, Carla, ed. 1987. Agricultural biotechnology. National Academy Press.

Colwell, R. (Forthcoming) Ecology and biotechnology: Expectations and outliers. In J. Fiksel and V. Covello (eds.) Risk analysis approaches for environmental releases of genetically engineered organisms. Springer-Verlag, Berlin.

Davis, B. 1987. Bacterial domestication: Underlying assumptions. Science 235:1329-1335.

Elfring, Chris, ed. 1987. New developments in biotechnology 1: Ownership of human tissues and cells. Office of Technology Assessment.

Forbes, Stephanie L., ed. 1988. New developments in biotechnology 3: Field-testing engineered organisms: Genetic and ecological issues. Office of Technology Assessment.

Forbes, Stephanie L., ed. 1988. New developments in biotechnology 4: U.S. investment in biotechnology. Office of Technology Assessment.

Forbes, Stephanie L., and Richard A. Danka, eds. 1987. New developments in biotechnology 2: Background paper: Public perceptions of biotechnology. Office of Technology Assessment.

Halvorson, Harlyn O., David Pramer, and Marvin Rogul, eds. 1985. Engineered organisms in the environment: Scientific issues. American Society for Micro-biology.

Introduction of recombinant DNA-engineered organisms into the environment: Key issues. 1987. National Academy Press, Washington, DC.

Mellon, M. 1988. biotechnology and the environment. National Wildlife Federation, Washington, DC.

Mellon, M. 1987. Biotechnology. In *S. Novick, D. Stever, and M. Mellon (eds.) Law of environmental protection, Vol. 2. Clark Boardman Company, New York, NY.*

Naj, A. K. 1989. Clouds gather over the biotech industry. Firms stumble on regulation, patent problems. Wall Street Journal, January 30; p. B1, B5.

Purchase, H. Graham, and D. R. MacKenzie, eds. 1990. Agricultural biotechnology: Introduction to field testing. Department of Information Services; Division of Agriculture, Forestry and Veterinary Medicine; Mississippi State University, MS.

Ratner, Mark. 1990. Survey and opinions: Barriers to field-testing genetically modified organisms. Bio/Technology 28:196-198.

Regal, P. 1987. Models of genetically engineered organisms and their ecological impact. Recombinant DNA Technical Bulletin. 10(3): 67-85.

Regulatory consideration: Genetically engineered plants. 1989. Center for Science Information, San Francisco, CA.

Sharples, F. 1987. Regulation of products from biotechnology. Science 235:1329-1332.

Tolin, S. A. and A. K. Vidaver. 1989. Guidelines and regulations for research with genetically modified organisms: A view from academe. Ann. Rev. Phytopathology 27:551-581.

CHAPTER SEVEN

The right of individuals to intellectual property protection on their inventions and writings is provided by the U.S. Constitution. These rights were extended to vegetatively reproduced plant materials in 1930 by the Plant Patent Act and to sexually reproduced plant varieties in 1970 by the Plant Variety Protection Act. Other biological materials—plant, animal, and microorganism—have been afforded protection under the utility patent law by a series of court decisions.

Utility patents provide broad protection for all kinds of intellectual property, including living materials. Genes, traits, DNA sequences, and related inventions can be patented so long as the patenting criteria of novelty, usefulness, and nonobviousness are met. Plant variety protection certificates, on the other hand, provide very narrow, variety-specific protection, subject to exemptions which allow researchers to use the protected variety for research and development and which allow farmers to save and sell seed of protected varieties. Secrecy can also provide protection for certain forms of intellectual property, provided critical, enabling information can be effectively withheld from the public.

The ability to legally protect intellectual property and secure exclusive use of an invention creates incentives for entrepreneurs to invest in the research and development needed to bring new products and services to market. Patents grant a period of exclusivity in exchange for public disclosure of the invention. Disclosure allows others to learn of the invention, advances the state of the art, and encourages additional inventions.

7 Intellectual Property Rights

V. L. Lechtenberg, Purdue University
A. A. Schmid, Michigan State University

Historical Perspective

The right of individual citizens to acquire and hold property is fundamental to the U.S. social and economic system. Private property ownership encourages risk-taking and entrepreneurism. The right of individuals to own intellectual property fosters rapid technological advance and inventiveness.

In the United States, the importance of such rights was recognized very early in our history. Article 1, Section 8, of the U.S. Constitution specifically empowers the Congress "to promote the progress of science and useful arts, by securing for limited times to authors and inventors the exclusive right to their respective writings and discoveries."

Pursuant to this section of the Constitution, the Congress of the United States passed the nation's first patent law in 1790. This law and related intellectual property laws are intended to foster and speed development and use of new inventions and new technology.

To achieve this goal, inventors are granted exclusive or monopolistic use of their invention for a finite period of time, as an incentive to disclose their invention to others. The patent holder is protected, during the period of exclusivity, from competition by others who might wish to make and sell products that embody the same invention.

Patents are not granted as a means of protecting secrets or withholding scientific and technical information from the public. On the contrary, patents constitute publication of pertinent information surrounding inventions. A valid patent must provide sufficient information that one of reasonable skill can duplicate the invention described by the patent.

Exclusive use is granted in exchange for disclosure in the belief that the public good is better served by disclosure than by secrecy. Once disclosed, others can learn from the invention, perhaps improve upon it, and generally advance the "state of the art."

The rights of intellectual property owners are similar to the rights of other property owners. They can sell or lease the intellectual property and enter into contractual and license arrangements with others regarding the property's use.

Intellectual property laws are found in nearly all technologically advanced countries. These laws differ among nations, and these differences can be the source of trade and marketing disputes. The U.S. and other nations have agreed to a number of international patent conventions and continue to strive for harmony among nations on policies for protection of intellectual property. Uniformity in patent policies, however, is yet to be achieved.

Evolution of Biological Property Rights

Utility patents have been the principal statutory basis for protection of

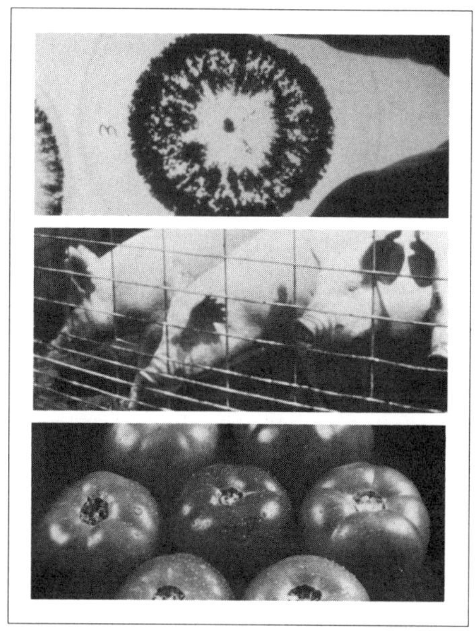

Living organisms were not considered patentable until 1980, when a Supreme Court decision extended protection to any biological material with uniqueness contributed by scientists.

intellectual property in the U.S. From the passage of the original 1790 patent act until 1980, however, it was held that living organisms were "products of nature" and not patentable under the utility patent law.

Congress provided alternative intellectual property protection for certain types of plant materials through passage of the Plant Patent Act of 1930 and the Plant Variety Protection Act of 1970. These laws addressed vegetatively and sexually reproduced plants, respectively, but did not provide protection for any other materials.

A shift in biological materials protection occurred in 1980 with a key U.S. Supreme Court decision. In the Chakrabarty case the Court ruled that any human invention, including a biological invention, was patentable under the utility patent law. This decision cleared the way for patent protection on living materials.

The Chakrabarty case involved the development of a genetically engineered bacteria with a unique trait. This "new" organism was clearly the result of human scientific intervention. The Court's decision, however, did not single out inventions created using specific techniques, such as genetic engineering techniques, nor did it limit patent protection to genetically engineered biological materials. Rather, it extended protection to any biological material with uniqueness contributed by scientists or inventors, without regard to the techniques used to achieve and perfect the uniqueness.

The Chakrabarty decision did not immediately result in patenting of plant and animal materials. The Patent and Trademark Office continued to hold that, in the case of plants, the Plant

Scientists differ on whether or not intellectual property developed through biotechnology should be treated differently with regard to patent protection. Some hold that it should, while others maintain that uniqueness is contributed, not by the research method, but by the imagination and creativity of the scientist.

Patent Act and Plant Variety Protection Act provided exclusive protection for inventors of plant materials. It was held that materials protected under either the Plant Patent Act or the Plant Variety Protection Act could not also be afforded utility patent protection. The justification for this belief was that, if Congress had intended to cover these materials under utility patent law, then the Congress would not have provided special protection afforded under the Plant Patent and Plant Variety Protection acts.

An important appeal decision (*Ex Parte Hibberd*) held that plants and parts of plants, including genes and fruits, could be patented under the utility patent law even though the plant material had previously been afforded protection under the Plant Variety Protection Act. This decision opened the door for protection of plant materials via utility patents. A related ruling (*Ex Parte Allen*) in 1987 afforded utility patent protection for animal inventions.

Since the Chakrabarty and Hibberd decisions there has been intense debate over the propriety of patent protection for biological materials. Some fear that patents on biological materials may interfere with scientific progress. Uncertainty over the long-term impacts of biotechnology and over the form in which products of this new biology will be produced and marketed has contributed to this fear.

Rights to Genetically Engineered Property

A number of scientists have suggested that, with regard to patent protection, intellectual property developed as a result of biotechnology should be treated differently than intellectual property developed using traditional research methods. While inventions derived using genetic engineering techniques are clearly the result of human intervention, it is not the research technique that contributes uniqueness to new plant or animal materials. Rather, it is the imagination and creativity of the geneticist and breeder in identifying specific genes, traits, and combinations of traits that contribute the uniqueness and utility necessary for patent protection.

Aside from protection of unique processes, per se, granting patent protection for unique living materials on the basis of the research techniques used fails to recognize the proper source of creativity. Genetic engineering and related technologies are simply tools at the disposal of the scientist. They complement other tools and techniques developed over decades of plant and animal research.

These new tools may make it possible to introduce into plants and animals traits that heretofore could not have been introduced, or they may simply make it easier to achieve results that could also have been achieved using older and traditional genetics and breeding procedures. Using these new tools speeds the process of plant and animal improvement; it does not necessarily make the products more creative.

In some situations, traditional breeding techniques or a combination of traditional and genetic engineering approaches may be superior to biotechnology for achieving certain plant and animal breeding goals. To base patenting decisions on whether or not certain techniques were used in research could create incentives for scientists to use less efficient procedures simply because patent protection could be achieved. Creating such incentives seems inconsistent with the intent of the patent law.

Forms of Protection

Biologically derived intellectual property can be protected through several approaches, depending on the nature of the property to be protected. The Plant Patent Act, the Plant Variety Protection Act, and the Utility Patent Act are all federal statutes that provide protection.

At the present time plant materials may be protected under one or more of these statutes. Microorganisms and animals can be statutorily protected only under utility patents. In addition to these federal statutes, intellectual property can also be protected through state laws governing contracts and trade secrets.

Plant Patents

The Plant Patent Act provides intellectual property protection for developers of vegetatively propagated plants. The holder of a plant patent can exclusively reproduce, sell, and use asexually propagated plants for 17 years.

Protection is not provided, however, for plant parts, genes, or traits. Also, there is no protection against sexual reproduction. Protection afforded under a plant patent is rather narrow and generally not applicable to protection of germ plasm. This form of protection is best suited to situations in which reproduction by seed is not possible.

Plant Variety Protection

Breeders of sexually reproduced plant varieties may seek protection under the Plant Variety Protection Act. The Plant Variety Protection Act is administered by the USDA, not the Patent and Trademark Office.

The critical criteria on which plant variety protection is based are novelty and sexual reproducibility. A protectable variety must be distinct from all others and genetically uniform and stable. A single trait or characteristic is sufficient to impart distinctness, regardless of the trait's usefulness.

Plant variety protection certificates provide patent-like protection to owners for a period of 18 years. Protection is very narrow and applies only to the specific variety or plant type described in the application for protection. Applications for variety protection certificates must be submitted within one year of the disclosure of the variety's existence. Seed deposits are required and assure that interested parties can acquire seed of the protected variety.

The Plant Variety Protection Act contains two key provisions that compromise its effectiveness as a protection instrument. The first is an exclusion allowing farmers to save seed of protected varieties for use in their own farming operations. This exclusion also allows farmers to sell seed of protected varieties to neighboring farmers for planting purposes, so long as income from seed sales is not the seller's primary source of income.

The intention of this provision is to allow farmers the freedom to use the seed they grow to satisfy their own future seed needs and those of their neighbors without infringing the rights of the plant variety protection certificate holder. The "sale to neighbors" provision, however, has led to abuses of the exemption. Some farmers have sold very large quantities of protected seed for seeding purposes to "neighbors" throughout the countryside. This abuse circumvents the owner's rights, has reduced the effectiveness of the Plant Variety Protection Act, and has created incentives for developers to protect varieties via utility patents.

The Plant Variety Protection Act also contains a research exemption which allows anyone to acquire and use seed of protected varieties for research and development purposes, without infringement. This includes the right to develop novel varieties from the protected variety, even if the derived variety is only cosmetically different from the original variety.

These new varieties can be sold without infringing the rights of the original plant variety protection certificate holder. The research exemption makes the Plant Variety Protection Act effective only for the narrow protection of specific varieties or lines. It is not effective for the protection of germ plasm, genes, or traits.

Some plant breeders argue that a single trait should not be sufficient basis for granting a protection certificate on an entity otherwise identical to an already protected variety. Some minimum, but as yet undefined, genetic difference or economic performance difference is suggested as an additional protection criterion. Such a criterion would prevent cosmetic breeding to circumvent existing protection certificates.

The single trait issue is complicated by the fact that some single-gene traits confer critically important agronomic characteristics. While these traits may be achieved by easily modifying an existing protected variety, the impact of the genetic change often is not trivial. The modified variety can represent a significant economic improvement over existing varieties.

Granting plant variety protection for these easily achieved but economically important genetic differences creates an incentive that spurs their development. Without this incentive it is doubtful that these varieties would be developed, at least not in the private sector. At the same time, current law creates an incentive for breeders to make cosmetic changes without improving plant performance.

The issue of how different a new variety must be to qualify for protection and avoid infringement is also an issue when materials are protected via utility patents. Resolution will depend on court interpretations.

Utility Patents

The long-established criteria for utility patents also apply to plants, animals, and microorganisms. These criteria are novelty, utility, and nonobviousness. Novelty means that the biological material is new and has not been previously described. Utility means that the invention has a useful purpose. Nonobviousness means that one reasonably skilled in the subject area would not recognize the invention as an obvious extension of the state of the art.

Utility patents provide broad protection. When applied to plants and animals, patents can provide protection for genes, traits, and DNA sequences as well as specific plant and animal products. Utility patents contain no provision specifically allowing farmers to save seed for future seeding purposes. Thus, what is perceived by many as a weakness of the Plant Variety Protection Act is overcome in principle by a utility patent on a plant variety. However, from a practical perspective, preventing farmers from saving seed may be difficult to enforce.

Criteria for Utility Patents

- *novelty*
- *utility*
- *nonobviousness*

Utility patents also lack the explicit research exemption found in certificates of plant variety protection. Many argue, however, that patent law allows use of patented material for research and experimentation so long as the end product of that research does not result in the sale of an infringing product.

While there is considerable debate over the research and experimental use issue, it is clear that patents are more restrictive regarding research and development use than certificates of plant variety protection. Certificates not only explicitly allow research to be conducted with protected materials, they allow sale of any derived variety without risk of infringement. Because patents do not allow such sale and provide the ability to protect genes and traits as well as products, utility patents provide the strongest form of statutory protection for unique germ plasm.

Trade Secrets

Some intellectual property is not suitable for protection under any of the statutes. In other cases the developer may simply elect not to seek statutory protection. Trade secret laws can provide protection in these situations. A trade secret can be any information that is not common knowledge and that gives its owner some form of business, technological, or trade advantage over competitors. The strength of the protection depends on the ability to keep critical enabling information secret.

Intellectual property owners wishing to protect their property through secrecy must actively take steps to maintain the secrecy. If the secret information becomes known, the protection is lost. Certain types of property are very poor candidates for trade secret protection simply because the enabling information is evident and disclosed by the product. Some biological materials fall into this category.

Some forms of biological materials are readily protectable via trade secret because of the inherent nature of the intellectual property. An example is hybrid corn. Because hybrids cannot be reproduced without access to the parent inbred lines, protecting the genetic material as a trade secret is not seriously compromised by the sale of the hybrid seed to farmers.

This form of protection has been sufficient to provide the incentives needed for research investments in the hybrid corn industry for many years. In addition, farmers are unable to grow their own seed without suffering significant yield losses. This form of biological protection, in concert with strict control of parent inbred lines, has provided excellent intellectual property protection.

Trade secrets may be suitable for protection of property in the private sector, assuming the critical information can be effectively withheld from the public. A good example is a microorganism which is kept in its creator's factory while only its product is sold.

Universities and other public institutions, however, will generally find trade secrets unacceptable as a means of protecting intellectual property. Universities must have the right to publish research results. Publications are critical to the progress of scientific research and are the primary means by which new knowledge is conveyed to other investigators.

Hybrid corn is an example of a biological material that is readily protectable by trade secrets.

While a reasonable delay in publication may be acceptable, if needed to secure patent or other protection, timely publication must be permitted. Publication requirements essentially preclude the trade secret approach as a viable protection mechanism for university-developed intellectual property.

Contracts

Agreements and contracts can be in the form of license agreements, sales contracts, and restrictive use provisions or labels. Combinations of more than one form may be used. With each approach the owner attempts to restrict or control the manner in which other parties may use the intellectual property.

The effectiveness of various contracts in controlling the unauthorized use of the intellectual property depends, to some extent, on the strength of the underlying property rights. A contract, backed by a patent or plant variety protection certificate, affords the best protection. Written agreements, signed by both parties, are the most desirable because they help assure that both parties understand and agree to the contract terms.

A weakness of contracts as the sole approach to protecting intellectual property is the fact that contracts are only binding on the contracting parties. Unless property rights are secured by patents or plant variety protection certificates, third parties who might acquire the property through legal means may not be restricted in its use.

A number of issues regarding the use of contracts to control biological property, in the absence of underlying statutory protection, remain unresolved.

Property Protection and Competitiveness

The number of patents and copyrights issued is frequently used as a measure of a country's inventiveness and technological competitiveness, although this measure does not capture qualitative differences. For many years the U.S. led the world in technological competitiveness. In recent years, however, a number of other countries have significantly increased patent filings and have become more technologically competitive.

This trend has encouraged the U.S. Congress and a number of state governments to provide funds and pass laws aimed at enhancing and speeding technological development. State governments have encouraged universities to undertake cooperative research and development programs with private companies. In the case of biological technologies, these efforts have been made easier by patent rulings of the last decade.

Several forces have been acting on both public and private institutions to stimulate interest in intellectual property rights. Beginning in the late 1970s U.S. industry, including agriculture, came under increasing pressure from foreign competitors. Preeminent positions in steel, automobile, and computer chip manufacturing eroded as we progressed into the 1980s. Agriculture has also faced unprecedented competition.

Several federal studies have addressed competitiveness issues. While these studies deal primarily with nonagricultural arenas, some of the findings also apply to agriculture, particularly as attempts are made to improve existing agricultural products, develop new products, and apply new technologies to crop and animal improvement. These studies have created a growing awareness that cutting-edge technology is critical to competitiveness and that future technological competitiveness will be based on "know-how" as well as on devices.

These studies also suggest that technology development and application in the U.S. can be improved. Interactions and joint research efforts among federal laboratories, universities, and private industry have been encouraged by the federal government.

The Stevenson-Wydler Act (1980) and the Federal Technology Transfer Act (1986) facilitated the transfer to industry of technology created in federal laboratories and made it easier for companies to work with federal scientists to develop new industrial products and services. State governments have also used their universities as instruments for state economic and high-technology business development.

Increasingly, technological competitiveness and successful high-technology-based economic development depend on the ability to protect new inventions and know-how through patents. If new biologically derived products that require significant development costs are to be perfected and marketed, then there must be appropriate mechanisms for protecting the underlying intellectual property.

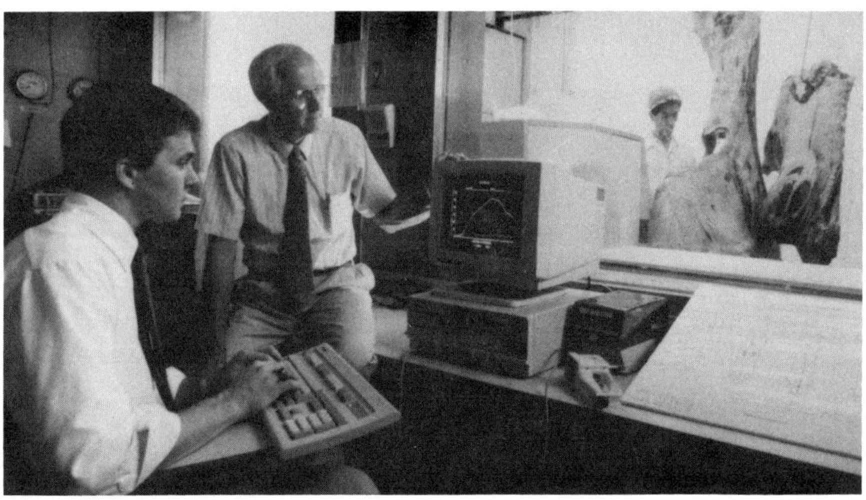

Technological competitiveness of U.S. industries depends on our ability to protect new inventions and know-how, particularly in the case of the new biologically derived products that require significant research and development costs for perfection and marketing.

Research Impacts of Intellectual Property Protection

The advent of utility patent protection for plant, animal, and microbial intellectual property has created uncertainty among scientists over exchange of research and experimental results. It is also not clear how patents will affect technology development and transfer of information from universities and federal laboratories to industry and to farmers and producers. This transformation of technology into products and services is essential to maintaining a competitive agriculture.

The passage of the Plant Variety Protection Act evoked similar discussion and apprehension among plant breeders and geneticists during the early 1970s. On balance, however, the impact of the act is generally perceived to have been positive. It is also important to remember that the utility patent law has generally worked well for 200 years. There is little or no evidence that it has impeded research and technology development. Most evidence is to the contrary. Technology transfer and adaptation have been enhanced.

Information Flow and Exchange of Research Materials

Some researchers and administrators, especially in public institutions, fear that the prospect of protecting intellectual property will create incentives for scientists to restrict information exchange, reduce research publications, and eliminate exchange of research materials. While there may be some risks, policies and procedures, such as use agreements, can be adopted that will mitigate these undesirable happenings.

It is also important to recognize that the absence of patent or other intellectual property protection does not ensure publication of research results and exchange of research materials. Some investigators and institutions, both public and private, have historically limited access to research materials, even in the absence of patents.

In fact, it is quite possible that researchers and institutions will be more willing to exchange materials and publish information when property is protected than when it is unprotected. This may be especially true if the property is perceived to have significant monetary value. Also, patents themselves, once issued, constitute publication of scientifically important information.

Use Agreements

Some scientists and investigators may choose not to exchange research materials with colleagues under any circumstances. In the long term, however, they will probably be disadvantaged because colleagues may reciprocate. A more likely scenario is that exchanges will occur under terms of written "materials use agreements." Such agreements are already common in many institutions and in the private sector.

University research administrators would be well advised to aid in developing these agreements, to assure that they conform to institutional policies. Their use can help assure that patents and other forms of protection do not unduly restrict publications or interfere with exchange of scientific information and research materials.

Use agreements that accompany transfer of research and breeding materials developed in the private sector to scientists in universities and federal laboratories should explicitly allow for use of the materials for research and educational purposes. For protection of the private sector, these agreements should prohibit transfer of materials to third parties without written permission to do so.

Use agreements should also contain appropriate language addressing ownership and potential licensing of any new intellectual property that might result from use of the transferred materials. The party receiving the materials should also agree not to sell any new products or processes derived from the exchanged materials without written permission to do so.

Similar agreements should govern private-sector use of materials developed in universities and federal laboratories. These agreements should generally grant a nonexclusive right to use the materials in research. Commercial exploitation, if permitted, should only be under terms of an appropriate license agreement.

International Exchanges

Plant and animal geneticists have historically searched the world for useful and novel traits. Some have argued that the advent of patents and intellectual property protection will curtail such searches. Others fear that poor countries will be exploited by having their natural germ plasm resources taken without payment while being forced to buy back improved varieties.

It can also be argued that intellectual property protection can create incentives to increase germ plasm exploration and conservation efforts. In this context, it is important to recognize that patents are issued for "inventions," not "discoveries."

Traits found in developing-country germ plasm must be improved and enhanced, through human intervention, into novel and valuable varieties or strains before patents and other forms of intellectual property protection can be obtained. Property protection can create incentives for the private sector to engage in this development.

Development of a protected or patented strain does not necessarily mean that the improved variety or strain will be well adapted, or even made available, in the country where the original germ plasm was found. In countries without intellectual property laws there is no incentive for the private sector to breed varieties uniquely suited to local conditions.

In fact, due to risk of theft, there may be significant incentives to actually keep out of these countries materials bred for markets in developed nations. Even in these situations, however, the utility of the original trait, in its original form, will not have been diminished. Its full potential, achievable through additional research and breeding, will simply not be realized.

The critical issue with regard to international implications of patents and property protection is not restricted access. Rather, it is the issue of how to (1) preserve genetic variety found in land races as commercialization of agriculture intensifies and (2) provide for the research funding necessary to use unique germ plasm in breeding varieties and strains well adapted in poor countries.

If viable markets for improved genetic materials do not exist and if these countries do not have effective protection for intellectual property, there is no incentive for the private sector. Public funding may be the only viable option. Public funds, however, are usually extremely scarce. Developed nations have recognized some obligation to assist in this process through support of the international research centers.

Issuance of a patent or plant variety protection certificate requires deposit of the protected materials in a form that enables others to duplicate the invention. These deposits may be in the form of seeds, cultures, clones, etc., so long as the materials are maintained in an enabling form. These deposited materials are available to requesting parties, including parties in other countries, and can be an important source of germ plasm for developing-country researchers.

The usefulness of deposited materials will depend on the level of technical skill and knowledge of the scientists requesting the materials. However, their availability eliminates, or at least greatly reduces, the risk that patents or plant variety protection can be used to keep others, including those in other countries, from acquiring access to new discoveries for experimental use. Obviously, those acquiring deposited materials must not use them in a manner that infringes on the patent or protection certificate.

Direction of the Research Agenda

One risk associated with intellectual property protection is that the public-sector research agenda may become skewed disproportionately toward product development and toward corporate research goals. Research administrators need to guard against this risk.

However, as the tools of biotechnology become integrated into traditional disciplines and as the appropriate role of intellectual property protection is established through experience, it is likely that the private sector will pressure the public research agenda toward more basic and fundamental investigations, the traditional role of the public sector.

The direction of the agricultural research agenda, in both the public and private sector, is determined by the availability of funds to support specific research programs. The exciting advances in biotechnology, along with opportunities for intellectual property protection, have created incentives on the part of companies to invest more heavily in biological research aimed at developing novel agricultural products and services.

Funds have become increasingly scarce in public institutions. Federal appropriations to agricultural experiment stations have diminished steadily, dropping almost 20 percent since 1980. The modest growth that has occurred in federal funding has been in specific research programs. Increases in state funding, where successful efforts have been mounted, have often been tied to specific programs also. Many have been biotechnology related.

These trends have diminished the ability of agricultural research directors in universities to fund meritorious programs in many fields of science. They have also caused research directors to look toward relationships with industry as an opportunity for securing additional research funding.

A survey by the Land-Grant Association's Division of Agriculture shows that scientific effort and expenditures for biotechnology research increased significantly during the 1980s. While some of this increase undoubtedly resulted from reallocation of resources from other scientific areas, much of the increase for biotechnology has been from funds that would not have been available for other purposes. These shifts also reflect response to emerging and promising science.

Public/Private Sector Relationships

The opportunity to protect plant and animal intellectual property has opened many opportunities for research cooperation between universities and companies. Each entity can more confidently approach the other with less risk that ideas and valuable technical information might be lost or compromised. A number of cooperative relationships have arisen as a result. Several technological competitiveness reviews conclude that these joint research relationships benefit both the university and the industrial partners.

Issues in University/Industry Agreements

In negotiating research agreements, it is important for each party to understand the goals of the other party. Universities usually wish to acquire funds to support research. They have the broad charge from society to engage in intellectual pursuits, advance the frontiers of knowledge, and teach students. To achieve these goals universities must be able to publish freely in the scientific literature.

Companies, on the other hand, wish to acquire access to protectable intellectual property in a manner that can lead to saleable products and services. Their overall goal is to recover their research investments in developing these new products and services and to make a profit. These goals are not necessarily incompatible. It is important to have institutional goals and policies clearly defined. The terms of the agreement will simply be the vehicle by which these policies are implemented.

Agreements covering university research supported by corporate funds must address issues related to access to intellectual property. These issues include the following.

Intellectual Property Ownership and Licenses

- research exemptions
- publication rights
- project control
- exclusivity of funding
- march-in rights
- confidential information
- project scope
- consulting arrangements

The Committee on Biotechnology of the Division of Agriculture, National Association of State Universities and Land-Grant Colleges has published excellent guidelines for university/industry research contracts. The Government-University-Industry Roundtable has also proposed model agreements. These model agreements and guidelines suggest appropriate language for the various terms in research agreements between companies and universities.

If sufficient care is taken in their formulation, such university/industry agreements can provide for the benefit and necessary protection of both partners. They also work to the ultimate benefit of the third party in all such arrangements: the public.

Benefits to Universities

Companies are often interested in basic research that they may not be able to conduct internally and in research that is not immediately applicable to a current product line. Companies and universities share many common research goals, such as gaining an understanding of the relationship between genes and gene products and understanding control of gene expression. Universities are often well suited to conduct this research if funding can be provided.

These sponsored research programs provide a benefit to the university by increasing the overall research program. This sponsorship also helps focus research on those problems with the greatest potential for real-world applications and significant economic impact. This focus is important in maintaining agricultural and industrial competitiveness.

A large percentage of graduate students trained in university research programs find employment in companies. The opportunity to interact with company scientists during their research training, as occurs with industry-sponsored research, improves understanding of the corporate research environment and establishes valuable job market contacts.

Benefits to Industry

Universities and federal laboratories are recognized as the primary performers of fundamental research. When technology is evolving rapidly, early access to the findings from these basic research programs is extremely valuable. Sponsorship of research provides companies an opportunity for this early access. University partnerships also provide companies an opportunity to establish contacts with scientific experts in fields that might otherwise be difficult to establish. These contacts can lead to important and productive research and consultancy relationships.

Companies also look to universities as the primary source of technically trained employees. Sponsored-research relationships offer an opportunity to become acquainted with undergraduate and graduate students, as well as postdoctorates.

Some companies, especially smaller and/or start-up companies, lack capacity for special research services, such as amino acid or DNA sequencing or DNA synthesis. Collaborations with a university can provide access to these services.

Benefits to the Public

University and industry research relationships and partnerships encourage economic development and enhance technological competitiveness. Successful relationships provide for appropriate protection of intellectual property, responsible licensing, and effective marketing. Patenting and licensing of our recently developed inventions is an effective method of technology transfer in high-technology businesses, including agricultural biotechnology. Some states have required that universities make provision for licensing and intellectual property protection as a prerequisite for acquiring special state research and development funds.

University and industry research relationships can aid in developing new businesses and speed the application of research findings to produce new products and services. Care must be taken, however, to balance the private-sector need to recover research and development costs with the public need to be fair in disseminating results and in providing equal access to publicly funded research.

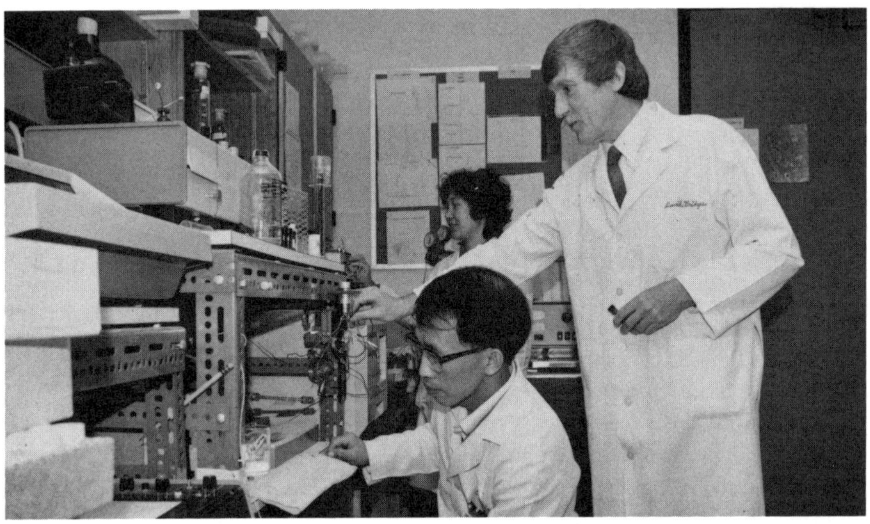

Joint research relationships between universities and companies can benefit both groups as well as society in general.

Some argue that providing the private sector access to technology acquired through cooperative university/industry undertakings is inappropriate in that licensees may acquire exclusive access to technology partially developed with public funds. Thus, the sponsor may benefit inappropriately from publicly supported research.

This position, however, must be weighed against the prospect that, without an opportunity for the licensee to recover development capital, the technology will go undeveloped. If this should happen, no one benefits from the public research investment, and society forgoes the economic gain that might be brought about by new business and new value-adding technologies. Limitations on the period of exclusivity, march-in rights, and reasonable royalty rates can all help assure that the public's interest is not abused.

Licensing of intellectual property to the private sector is likely to generate royalty income for universities. Most universities have long-established procedures addressing royalty rates and distribution to inventors. These procedures will most likely apply to biological as well as mechanical inventions. Royalty distribution procedures usually allow a percentage of the royalties to accrue to the inventors and part to be returned to the inventor's department to support additional research. This provides strong incentives to spur inventiveness, even in public institutions.

While the benefits and costs of protecting intellectual property in biotechnology fields must be constantly weighed, the opportunity for strong statutory protection has been positive. Research investments have been increased, leading to a better understanding of plant and animal systems. Dialogue and interaction among public- and private-sector scientists has been increased. This increase in research and greater interactions among scientists will rebound to the gain of society.

References and Additional Readings

Committee on Biotechnology, Division of Agriculture. 1984. Guidelines for development of university/industry research contract. p. 50-65. In *Emerging biotechnologies in agriculture: Issues and policies. Progress Report III. National Association of State Universities and Land-Grant Colleges, Washington, DC.*

Butler, L. J. and B. Marion. 1983. Economic impacts of Plant Variety Protection Act. North Central Project 117 Monograph, Madison, WI.

Government-Industry-University Roundtable. 1988. Simplified and standardized model agreements for university-industry cooperative research. Industrial Research Institute. National Academy Press, Washington, DC.

Ihnen, Jeffrey L. and Robert J. Jondle. 1989. Protecting plant germ plasm: Alternatives to patent and plant variety protection. In *Intellectual property rights associated with plants. Spec. Pub. 52. American Society of Agronomy, Madison, WI.*

Kloppenberg, Jack R., Jr., ed. 1989. Seeds and sovereignty: The use and control of plant genetic resources. Duke University Press, Durham, NC.

Lechtenberg, Victor L. 1989. Intellectual property rights—Industry/university agreements. In *Intellectual property rights associated with plants. Spec. Pub. 52. American Society of Agronomy, Madison, WI.*

Office of Technology Assessment. 1984. Congress of U.S. commercial biotechnology: An international analysis. U.S. Government Printing Office, Washington, DC.

Schmid, A. Allan. 1985. Biotechnology, plant variety protection, and changing property institutions in agriculture. North Central Journal of Agricultural Economics 7(2): 129-138.

Stallmann, Judith I. and A. Allan Schmid. 1987. Property rights in plants: Implications for biotechnology research and extension. American Journal of Agricultural Economics 69(2): 432-437.

Williams, Sidney B. and Kenneth A. Weber. 1989. Intellectual property protection and plants. In *Intellectual property rights associated with plants. Spec. Pub. 52. American Society of Agronomy, Madison, WI.*

CHAPTER EIGHT

Biotechnology appears to be the key to further productivity gains in agriculture into the next century. These gains are necessary to ensure adequate and nutritious food supplies worldwide, and they are essential to the continuing economic vitality of developing countries.

However, the shift in applied research and associated product development from the public to the private sector suggests that benefits from advances in applied research will become less widely available. And there is little likelihood that the products developed will be ones which are important to poor, developing countries. What, then, are the likely international impacts of biotechnology advances, and what can be done to ensure that poorer countries share in the benefits?

Advances in biotechnology will certainly affect the patterns of world trade. Without efforts to liberalize trade and to aid developing countries in conducting applied research, developing countries—the main markets for developed-country exports—may experience economic deterioration and decreased ability to import. Severe dislocations in agriculture may then occur in developed countries, as well.

Biotechnology is knowledge intensive, and its applications tend to be location specific. Many developing countries have no basic research capacity, limited capabilities to adapt biotechnology advances to local conditions, and few resources to attract private transnational corporations. For the continued well-being of an increasingly interdependent global economy, a conscious effort must be made to help developing countries acquire the appropriate technology and establish and maintain an infrastructure to support applied research. International Agricultural Research Centers and similar, public-oriented institutions could provide such assistance while also ensuring that biosafety measures, intellectual property rights, and legal regulations are observed.

8 Agricultural Biotechnology: A Global Perspective

Randolph Barker, Cornell University
Don Plucknett, The World Bank-CGIAR

Introduction

Since World War II, technological change has been recognized as an increasingly important source of output and productivity growth in both developed and developing countries. How then does biotechnology differ from earlier forms of technological change? How will biotechnology affect the competitiveness of U.S. agriculture? How will it impact on world trade? What are the implications for Third World development? These are the questions addressed in this chapter.

Over the past half century, world agriculture has been undergoing a remarkable transition. Because there will be few new lands to plow by the end of this century, most additional agricultural output will have to come from intensive use of previously cultivated areas.

Today, over 80 percent of output growth in world agriculture comes from higher yields. This productivity growth is dependent in large measure on petrochemicals and fossil fuels. With the maturing of the petrochemical and transportation industries, the potential for further growth in productivity from these sources has diminished. Furthermore, the rapid rise of energy consumption, heavy dependence on fossil fuels, and increasing reliance on agricultural chemicals have led to a growing concern about resource depletion and the pollution of the environment.

Biotechnology and microelectronics (or information technology) are the new core technologies which offer the prospects for further productivity growth, savings in raw materials, reduction in energy consumption, and a lessening of pollution problems. These two technologies are becoming increasingly interlinked. Biotechnology is knowledge intensive, and advances in biotechnology require considerable scientific manpower. Access to this knowledge and to the economic power that it represents has become a key political issue.

Biotechnology is not a substitute for but a complement to other forms of technology. In some, but not in all, cases, biotechnology development may enhance the speed or lower the cost with which we are able to achieve certain goals such as increasing the yield or improving the quality of a commodity. In other cases, such as biological nitrogen fixation, biotechnology may make it possible to achieve goals that cannot be achieved with conventional methods.

There are many ways in which biotechnology is likely to affect agriculture:

(1) through increased yields and productivity;

(2) through improved quality and shelf life of products;

(3) through improved resistance to pests and diseases;

(4) through the substitution of one crop for another or, alternatively, a synthetic for an agricultural commodity;

(5) through the substitution of renewable (agricultural) for nonrenewable (fossil fuel) energy sources;

(6) through the adaptation of existing crops to different environments; and

(7) through the shift in marketable surplus (the proportion of agricultural output in excess of that needed for home consumption) to farmers, to regions, and to nations with high productivity.

The authors wish to thank Lowell Hardin, Vernon Ruttan, and Martha Bair Steinbock for useful comments and suggestions.

All of the above shifts were underway to a greater or lesser extent even before the advent of biotechnology. The changes taking place are evolutionary. Gains in agricultural productivity in the next decade or two will come largely from the spread of existing technologies. However, the growing dependence on science-based agriculture and associated advances in biotechnology and information technology are likely to speed up the process.

When we look back 20 or 30 years from now, the changes very likely will seem to have been revolutionary. Access to knowledge and the ability to adapt institutions to meet new circumstances will be crucial both (a) for the developed countries attempting to maintain a competitive position in world markets and (b) for the developing countries hoping to gain access to new technologies to enhance the productivity of their own agricultures.

The first section that follows describes the change in the institutional structure of research that is occurring. What is different about biotechnological advances relative to previous technological changes is the ability to patent these new technologies. Because the private sector can profit from biotechnology research, the roles of the public and private sector are changing.

The next two sections deal with the implications of advances in biotechnology for the competitiveness of U.S. agriculture and for world trade. For the future well-being of all, openness in science, technology exchange, and trade needs to be fostered and sustained.

Finally, the last major section discusses the implications for technology transfer and agricultural development in the Third World. Even if special efforts are made, the advent of biotechnology is likely to widen rather than narrow the gap in productivity between the industrialized and developing countries.

The Changing Institutional Structure of Agricultural Research

Historically, gains in agricultural productivity have depended heavily on public investment in both basic and applied research. Except in the case of hybrid seeds, the private sector has rarely captured the benefits from the development of technologies such as improved seeds and cultural practices. For almost a century, beginning with the Morrill Act of 1862, the bulk of agricultural research in the United States was conducted in state land-grant colleges and experiment stations and disseminated to farmers through the state-federal cooperative extension service. This highly successful, publicly supported system is decentralized and articulated. It facilitates a flow of ideas among farmers, extension workers, and researchers.

By the 1970s, however, the land-grant research system came under increasing criticism. One argument was that the system had become a publicly subsidized research arm of agribusiness and the large farmer. A second criticism was that agricultural research was becoming highly insular and largely divorced from the frontiers of knowledge in the basic biological sciences. With the advent of biotechnology have come changes in government and private-sector support for and in the institutional structure of research. These changes have addressed the concerns about basic research but not the charge that the system is the tool of agribusiness and the operators of large-scale farms.

What has happened is this. The emergence of biotechnology in the 1970s led to a series of judicial interpretations relating to intellectual property. These interpretations make it possible (for private-sector and public-sector institutions) to obtain patent protection for living plant and animal materials. This has led to a new division of labor in research between public and private sectors. Increasingly, the public sector

For almost a century, the bulk of U.S. agricultural research was conducted at state land-grant colleges and experiment stations, but much of the agricultural biotechnology research is being done elsewhere.

is focusing on basic research, while the private sector is concentrating on technology development and diffusion that is patentable. The private sector recognizes, however, that its own progress in biotechnology applications depends on the rate of progress made in basic research, most of which is publicly supported.

Many of the researchers in this new alliance do not belong to the traditional land-grant establishment. In fact, much of the research in agricultural biotechnology has moved outside of the land-grant colleges and experiment stations to basic science laboratories in non-agricultural colleges and to basic and applied research laboratories in private-sector companies.

These changes have been accompanied by a shift in the sources of funding for research even within the public sector. There has been a sharp drop in "formula funding" over which the land-grant colleges exercised control and more direct support to individual scientists and laboratories. This comes about primarily through grants from the National Institutes of Health, the National Science Foundation, and from private companies.

Research on biotechnology represents about 10 percent of total expenditures on agricultural research, with half of the total in the public and half in the private sector. However, modern biotechnology research in agriculture is being driven by private-sector demands. The research-extension-farmer linkage has been weakened, and many state land-grant colleges are continuing to be criticized for not serving the interests of the "small farmers." This criticism, in some instances, has encouraged the agricultural colleges to concentrate on more basic research.

However, with biotechnology it is difficult to draw sharp lines between basic research and technology development. It is also hard, if not inaccurate, to argue that basic research is "neutral" or unbiased, or to infer that most scientific discoveries are serendipitous, obviating the need for public institutions to set research priorities. A continuing important role exists for the public sector to develop and disseminate technologies with a high social payoff, technologies that the private sector will not develop because it is not profitable for it to do so.

In some instances it may be necessary to create new institutions and organizations to insure that public-sector interests are addressed. Two examples of such institutional innovations that have already been made are the Michigan Biotechnology Institute and the International Program on Rice Biotechnology, supported by the Rockefeller Foundation.

The Michigan initiative is an independent, nonprofit research and development corporation designed to foster the development of biotechnologies important to the state. The Rockefeller Foundation, in deciding to focus on rice, identified an important need for the developing countries that was not being addressed. In response, the foundation funded a worldwide network of approximately 50 laboratories in the developed countries to conduct basic research on rice. Through collaboration with the International Rice Research Institute in the Philippines, this research is being linked directly to the needs of the developing countries.

To what degree have America's main competitors in science and technology development, Japan and Europe, followed the U.S. pattern in the development of biotechnologies? First, the U.S. investment in basic research in biotechnology totally overshadows the basic research investments by Japan and major European countries combined. In 1984, for example, U.S. federal government funding for biotechnology research was approximately $500 million, more than double the amount spent by West Germany, France, Great Britain, and Japan combined.

Three Biotechnology Research Fields Selected for Subsidized Government Funding in Japan

- *fermentation*
- *large-scale cell culture*
- *recombinant DNA*

Second, and even more important, the configuration of institutions—government, universities, and industry—as well as the pattern of government intervention differs among countries. The American university system, where most of this research is conducted, has no counterpart in basic research capacity elsewhere in the world. While the structure of Japanese industry serves to constrain the creation of venture capital firms, there has been a long history of close cooperation between government, universities, and industry in applied agricultural research in that country. A nation's or a firm's competitive advantage depends not simply on its basic research capacity but also on its ability to develop and market biotechnology applications. The significant institutional differences among countries will affect the strategies they choose and policies they adopt in developing biotechnology capacity to enable them to remain competitive in world markets. We expect Japan to be at the forefront in production and marketing of biotechnology at the applied level just as it is in electronics.

Implications for the Competitiveness of U.S. Agriculture

We begin this section with a discussion of the determinants of international competitiveness. Then we comment on protectionism and on other policies being adopted by many of the developed countries that may affect their competitive position.

Determinants of International Competitiveness

A three-step process of invention, innovation, and diffusion is involved in the development and application of technology. It is difficult to draw sharp lines between these areas because productivity increases in agriculture through biotechnology necessitate close links among all three.

Due to the scope and magnitude of the U.S. research and industrial establishment, it appears at first glance that the United States would have a comparative advantage over its competitors in exploiting biotechnology. However, in the "technology war" the bottom line is market share. In the microelectronics field, there is already concern over the erosion of America's market share as the Japanese exploit their advantage in applied engineering. Is this about to happen in biotechnology, as well? (Much of this section is based on Marin Fransman, 1988. Biotechnology - generation, diffusion, and policy: An interpretative study. Mimeo, University of Edinburgh.)

What the Japanese experience in microelectronics illustrates is that one need not have strength in basic research in order to be competitive in world markets. There is already concern in the United States that competitors are targeting biotechnology industries to receive government support and that these industries are drawing heavily on U.S. basic research knowledge. For example, the Japanese government has been targeting assistance to agricultural biotechnology research and development. However, Europe, Canada, and Australia have more research interest than Japan in the agricultural commodities that are of direct concern to the United States, including dairy and major cereal grains.

Beyond research investments as such, the size and configuration of firms, the degree of cooperation among firms, and the extent of collaboration between the private and public sectors have an important bearing on international competitiveness. In the U.S. the number of venture capital firms devoted to biotechnology research mushroomed early on. Now most of those that have survived have had to negotiate agreements with larger companies to take advantage of marketing facilities.

In Japan beginning in 1984, the Ministry of International Trade and Development (MITI) identified biotechnology as one of the technologies to be included in the "next generation basic technologies development programme." Three research fields were selected for subsidized government funding: fermentation (where Japan has had traditional strength); large-scale cell culture; and recombinant DNA. The MITI has tried to encourage research collaboration among firms by involving the 14 largest ones in research associations focused on the three priority areas. U.S. antitrust laws would prohibit such collusion here, but the degree to which government policy has fostered effective collaborative research in Japan is not clear.

The experience in both the United States and Japan suggests that the economies of scale may be such that the control of major segments of biotechnology development will fall into the hands of a few large firms. Thus, only a handful of the economically most advanced countries are in a position to develop biotechnologies for export.

While the majority of countries have the capacity to adapt biotechnologies to local conditions, the poorest countries lack even this capability. Efforts to perfect tissue culture and other techniques that will facilitate biotechnology transfer are already underway in the International Agricultural Research Centers (IARCs) and many of the National Agricultural Research Centers (NARCs) in the developing countries.

Protectionism

The rapid growth in world agricultural trade and U.S. exports in the 1970s was associated with a significant increase in the share of U.S. exports that went to developing countries (Table 1). The 1980s witnessed a slowdown in economic growth and world trade in both the developed and the developing countries. Agricultural commodity prices fell and protectionism increased as countries sought to insulate their producers from declining world prices. A 1986 editorial in the Economist described the situation as follows:

> In the rich and mainly industrial countries farmers are paid too much, so they produce too much. In the poor and mainly agricultural countries farmers are paid too little, so they produce too little. Europeans trample Cognac grapes into industrial alcohol; Americans fill Rocky Mountain caverns with butter; Japanese pay eight to ten times the world price for their bowl of rice. Meanwhile, many millions of Asians and Africans live in rural poverty and go hungry to bed. Do not despair. The mistakes are so large that these countries' policies will soon collapse. Properly staged and handled, that collapse will leave the whole world better off (Economist, November 15, 1986, p. 13.)

The author of this editorial is more sanguine about the prospects for trade liberalization than seems to be warranted by the slow progress toward major reforms in the current General Agreement on Tariffs and Trade (GATT) negotiations. The question of why such seemingly irrational levels of subsidy have occurred in almost all developed countries is seldom examined. Trade liberalization would benefit consumers but heavily penalize farmers, particularly in Europe and Japan. However, there must be strong political pressures for maintaining protectionist policies which transcend rural interests. While these are not well understood, they provide consistent support for continuing high levels of subsidies.

The United States was instrumental in promoting measures to exempt agriculture from earlier trade liberalization measures in GATT. Now the United States is arguing strongly for the liberalization of trade in agriculture, but it is simultaneously arguing for a strengthening of protection for intellectual property rights, and in some quarters is promoting measures to restrict the flow of scientific information. The rational for the position of protectionists is this. A handful of developed countries have a strong lead in biotechnology research. Limiting the flow of information, these countries feel, should enhance their competitive position in world markets. It is thought that the United States, with the largest investment in biotechnology research, stands to benefit most if protectionist measures can be successfully implemented.

However, the impact of protectionist policies on agricultural development and world trade is not as clear as the arguments of the protagonists suggest. One of the most controversial pieces of legislation, the so-called "Bumpers Amendment" (named after Senator Dale Bumpers), was passed into law in May 1986 (U.S. Senate, 1986). In Bumpers' words, the act is to "prevent American tax dollars from being used to help foreign countries who are trying to take

Table 1. *Distribution of U.S. Agricultural Exports Among Developing, Developed, and Centrally Planned Economies.*

Year	Total (In billions of dollars)	Percentage of total exports to: Developing economies	Developed economies	Centrally planned economies
			All agricultural exports	
1970	7.4	33.0	65.1	1.9
1985	30.2	42.2	49.2	7.8
			Food grains	
1970	1.4	65.7	33.7	0.7
1985	4.5	72.0	20.4	6.1
			Feed grains	
1970	0.9	6.8	92.2	0.9
1985	5.4	28.6	38.9	29.9
			Oilseeds and products	
1970	1.4	19.6	79.1	1.2
1985	4.4	31.5	66.1	1.7

Source: Congressional Budget Office. 1989. Agricultural Progress in the Third World and Its Effect on U.S. Farm Exports, *Based on data from Department of Agriculture, Economic Research Service.*

our export markets." This law prohibits the U.S. Agency for International Development (USAID) from lending assistance to foreign institutions to be used for conducting research on commodities which would compete with U.S. exports.

On the other hand, opponents of such legislation argue that restricting support for research slows agricultural and economic growth in developing countries and hence slows growth in their demand for agricultural exports. Evidence to support this argument is shown in Figure 1. Over time, as incomes have risen in the more advanced lower, and upper-middle-income developing countries, the level of self-sufficiency in total grains has been declining. The resulting increase in import demand is principally for coarse or feed grains for which the U.S. is a major supplier.

The impact of strengthening of intellectual property rights is also controversial. Such action affects different groups of countries differently. Those countries with the capacity for local adaptive research want foreign firms to provide inventions but want them to do so at low cost. Even so, because such countries receive nothing in return for protecting the intellectual property rights of foreign firms, there is a widespread tendency to acquire technology through noncompliance with legal controls or to "pirate." The poor countries can gain no advantage from pirating, since they need assistance at all levels of technology transfer. But, as a rule, private foreign firms are not attracted to these countries. As a whole, therefore, the developing countries have limited capacity to deal with intellectual property rights. Further, neither the IARCs nor the NARCs have given this matter much attention. Simply put, it is profitable for the developing countries to avoid committing themselves to international patent and copyright conventions and to "beg, borrow, or steal" biotechnology from the developed countries.

In theory, the strengthening of intellectual property rights could increase the competitive position of U.S. agriculture by encouraging greater private-sector investment, which makes newer technologies available at lower cost. However, it is difficult to judge how effective efforts to restrict the flow of new knowledge and new technology really are. Furthermore, such efforts could be counterproductive if they dampen the growth of agriculture in Third World countries. There is clear evidence that, as their incomes rise, these countries increase their imports from the developed countries.

Figure 1. *Self-sufficiency Ratios for Grains by Developing Country Income Groups (1.0 equals self- sufficiency).*

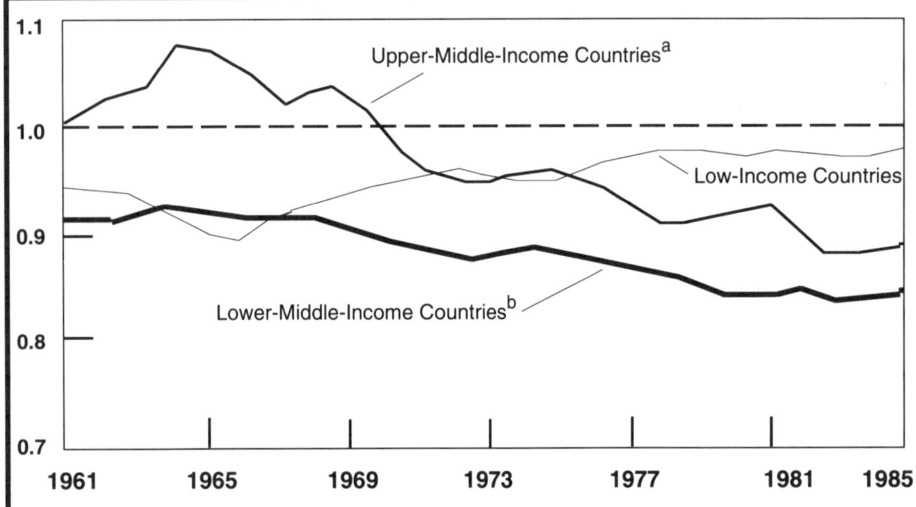

Note: The self-sufficiency ratio equals domestic grain production divided by the sum of grain production plus net imports. A country with no imports or exports has a self-sufficiency ratio of 1.0.

[a]. *Excluding European countries and high-income oil-exporting countries.*

[b]. *Excluding China and other Asian centrally planned economies.*

Source: Congressional Budget Office, based on Gary Vocke, "U.S. Grain Imports by Developing Countries," Department of Agriculture, Issues in Agricultural Policy, Number 542 (May 1988)

Concerns About Health and Environmental Hazards

The public concerns about health and environmental hazards associated with the introduction of new biotechnologies could have a major impact on the introduction or acceptance of certain biotechnologies, particularly in the industrial countries. This in turn could affect the competitive position among countries for specific commodities, if some countries approved a given

Table 2. *Trade in Farm Products.*

Product group	Percent Export Share (1986) Developed countries	Developing areas	Eastern bloc
Dairy products	95.2	1.7	2.9
Beverages excluding tropical	87.9	5.2	6.9
Pulp	87.7	7.1	5.2
Hides and skins	86.2	6.9	5.7
Cereals and cereal preparation	82.5	12.1	5.7
Meat and meat preparation	78.2	11.4	10.3
Tobacco and tobacco products	70.2	21.3	8.5
Crude materials	68.5	22.5	9.0
Wood	65.3	21.9	14.1
Fruits and vegetables	61.5	30.7	7.2
Oilseeds, fats, oils, oilcakes, etc.	61.4	32.4	6.4
Fish and fish preparation	57.7	35.8	6.5
Natural fibres	48.7	30.9	20.4
Sugar	23.3	71.3	4.0
Tropical beverages	9.4	88.7	1.9
Natural rubber	2.6	96.3	1.5

Note: Figures have been rounded.
Source: GATT.

technology and others restricted its use. This would be somewhat akin to the effect of pollution-control regulations on the location of certain industrial firms. Public concerns are being reflected in the regulatory and trade policies being adopted by the Environmental Protection Agency (EPA) and other government regulatory agencies and in those being formulated by the European Community in preparation for integration in 1992.

The potential impact of social policy and public perceptions on biotechnology is illustrated in the case of bovine somatotropin (BST). There have already been restricted tests of this growth hormone in commercial dairy herds, and economists have predicted that the increase in milk yield per cow will lead to lower cost of production and ultimately to lower milk prices for consumers. But there is concern among some consumer advocate groups that milk from cows injected with BST could pose a health hazard. Although there is no evidence to substantiate this, surveys do show a perception of a health risk among a significant portion of the public. What impact this will have on the timing or delay in approval and introduction of BST or on the consumption of milk if BST is approved for use is hard to predict.

In summary, public concerns regarding health or environmental hazards could block or delay the introduction of certain biotechnologies. This in turn could affect the competitive position of countries in those affected commodities. This might lessen the competitive advantage that biotechnology seems to offer the U.S. and other industrialized countries over the developing countries.

Impact on World Trade

Agricultural trade is extremely important in the development process. As noted previously, as incomes rise, developing countries often experience difficulty in producing enough food, such as cereals and oils, as well as other products needed to promote industrial development (Figure 1). By contrast, developed-country subsidies bring forth a surplus of agricultural products for export. Not surprisingly, therefore, the demand for U.S. agricultural exports is growing much more rapidly in the developing than in the developed countries (Table 1).

Table 2 shows the percent share of export markets for developed, developing, and eastern bloc countries by major product groups. Developing countries have long been associated with the export of raw materials, and very often exports are concentrated in one or two major commodities such as sugar, coconuts, palm oil, cacao, coffee, tea, and rubber. By contrast, food grains, feed grains, and oilseed products account for 50 percent of all U.S. exports.

One of the important impacts that biotechnology will have on world trade is through the large-scale substitution of certain agricultural commodities for other commodities. Biotechnology makes it possible to focus on the utilization of end-products such as starches, proteins, cooking oils, or sweeteners. Thus, biotechnology can drastically change the comparative advantage of particular crops in the production of any of these end-products. The effects of such advances on crop production and exports are illustrated by what has happened with sweeteners and vegetable oils.

The enzymatic digestion of corn starch to obtain high-fructose corn syrup (HFCS) was first introduced into commercial production in Japan in 1967 and spread to the U.S. in 1972. Today, liquid corn sweetener has replaced approximately 50 percent of the sugar consumed in the United States and accounts for 15 percent of the world sweetener market. This new product has spread most rapidly in Japan and the United States, where high price supports have been maintained to protect domestic sugar beet industries. However, the use of liquid corn sweetener is still restricted in Europe, where biotechnology is being used to improve the production of sugar beets and to develop alternative sweeteners.

The development of sweetener substitutes led to a sharp decline in world sugar exports. U.S. sugar imports dropped from 4.6 million tons in 1978 to 2 million tons in 1986. One of the most severely affected countries was the Philippines. In the early 1970s under a preferential tariff agreement with the United States, the Philippines exported over 1.3 million metric tons of sugar to the United States. Sugar was the second largest foreign exchange earner. By 1986 Philippine sugar exports had fallen to about one-fifth of the previous level.

What is the potential for sugar in the future? The application of biotechnologies in the production of substitute sweeteners is likely to spread to developing countries that are now sugar importers. The trend will be toward self-sufficiency in the production of sweeteners. However, a new biotechnology process patented under the name of SUCROTECH is now being commercially marketed. Ironically, this new approach to making ethanol also has the potential to produce fructose from sugarcane more cheaply than from corn. However, it seems highly unlikely that world cane growers will regain their lost export markets.

The major deficit regions for oils and fats are Asia, the Middle East, Africa, Latin America, and the USSR. Soybean and oil palm have become the two most important sources of vegetable oils and fats, with oil palm having overtaken copra (coconut) in the 1970s. However, from 1983 to 1987 the production and consumption of rapeseed oil had the fastest rate of growth (9.9% per annum for rapeseed, 8.5% for palm, and 2.8% for soybean). The rapid growth in oil palm and rapeseed reflects significant technological advances in both crops.

Over 80 percent of world oil palm exports come from one country, Malaysia. Through cooperation with Unilever, a Dutch multinational firm, the Malaysian oil palm industry has been able to develop genetically improved palms that have rapidly increased yields. The first application of biotechnology to this crop has been in the micropropagation of clonal palms from selected elite individuals. Tissue culture propagation of oil palm has been researched for two decades. This technique offers the potential for increasing yields by 30 percent. However, commercial application has been delayed by the occurrence of flowering abnormalities in some of the first clones.

Despite the lack of success to date with tissue culture, Malaysia has been able to dominate the export market, and this has had a significant effect on other developing countries. Palm oil exports from West Africa, once a leading exporter to Europe, have continued to decline. The Philippines has also suffered adversely from the expansion of palm oil production, in

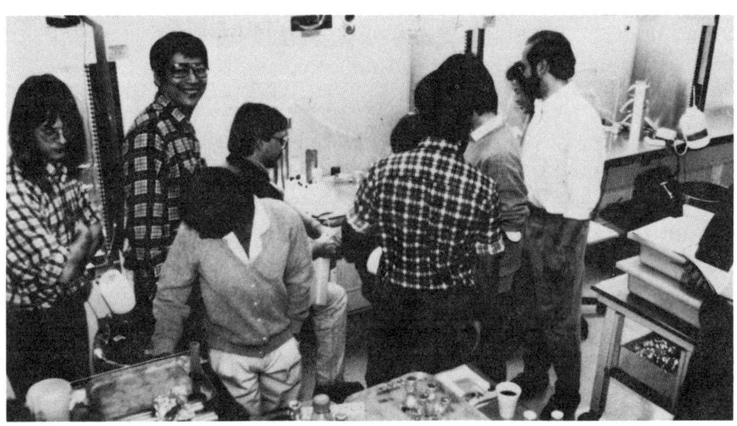

The United States helps transfer technology through scientific research and training, direct trade, and agricultural development.

part because its own research on coconuts has not kept pace. Over the decade that ended with the ouster of former President Marcos, the Philippine government squandered large sums of money ostensibly collected from coconut and sugarcane producers for investment in research.

What does the future hold for palm oil exports? Successful commercialization of cloning would be a major boost to producing countries. However, much will also depend on technical advances in competing crops. Rapeseed is a significant source of vegetable oil in many of the deficit areas of the world. Biotechnology research led to Canada's development of canola, a superior-quality rapeseed with low erucic acid in the oil. Canola is rapidly becoming the international standard for trading purposes. The prospects of improving rapeseed yields through biotechnology are very promising.

Sweeteners and vegetable oils are but two examples of how improvements in either the quality or the yield potential of a crop can have a significant effect on world export markets. These examples suggest that widespread application of biotechnology could increase the volatility of world export markets and lead to dislocations in agriculture in both the developed and the developing countries.

The protectionist policies of the developed countries will provide greater incentives for researchers to provide substitutes for imports, including those from developing countries. The developing countries will in turn search for technologies which will enable them to reduce their imports. The overall effect could be to encourage a higher degree of agricultural self-sufficiency, but this would be counter to historical trends.

Implications for the Developing Countries

Countries can be classified into three categories with respect to their capacity for biotechnology development and utilization: (1) those with the capacity to export biotechnologies, (2) those with the potential to adapt biotechnologies to local conditions, and (3) those with virtually no applied research capacity. With the possible exception of a handful of Newly Industrialized Countries (NICs), the developing countries fall into one of the latter two categories.

New biotechnologies developed in the first group of countries will become an increasingly important potential source of productivity growth in the developing countries. But the developing countries face a number of problems. How do they go about adapting these new technologies to their conditions? How can they ensure that at least a small portion of the basic and strategic research done in the laboratories in the developed countries focuses on commodities or problems of particular importance to Third World countries?

Technology Transfer

The first question asked above is normally discussed under the heading of technology transfer. Because of the biological nature of agriculture, technologies tend to be very location specific. There is a tendency to underestimate the difficulties involved in agricultural technology transfer and the development of appropriate, site-specific technology.

A considerable portion of technology transfer is in the form of scientific information, knowledge, and techniques. The United States, as one of the leaders in most aspects of agricultural technology, helps transfer technology through direct trade, scientific research and training, and agricultural development programs. The pace of technology transfer has increased over the past two or three decades. The world has enjoyed not only a heavy public investment in the development of new biological technologies, but also a high degree of public control over extension processes and institutions which dispense the technology free to farmers.

With biotechnology development, however, the private sector is playing a much more important role. Concern is mounting that the private-sector research agenda will be too narrowly focused on the well-to-do farmers (those who have the capacity to purchase inputs) and on the well-to-do countries that have good agricultural support services. Among the more than 60 developing countries, perhaps only a dozen would provide attractive markets for transnational companies.

Further conflict between public- and private-sector interests is developing in the area of cereal grain research. For example, in 1980 the Chinese signed an agreement with Ringaround Seed Products (Occidental Petroleum) and Cargill giving them the exclusive rights to the production and marketing of hybrid rice in certain areas of the world. Since that time, these private companies have pressured the Chinese not to share the information and germplasm involved in their hybrids with the International Rice Research Institute (IRRI). In another case, rather than accept the restrictions imposed by a private company on the use of its research in wheat gene mapping, the International Maize and Wheat

Improvement Center (CIMMYT) has sought public-sector funding to support wheat gene mapping research in the United States.

Despite the unresolved controversy generated by the assumed tendency of the Green Revolution to serve narrow private ends, it is important to remember that this set of major advances was conceived and implemented within the public sector. If, in a like manner, biotechnology is to serve the interests of developing-country farmers, public-sector institutions must continue to play a central role in technology development and diffusion.

Developing countries can acquire new technologies in several ways. They can either make, buy, or pirate them. National programs may work directly with the private sector through the international centers or through linkages to public-sector institutions in the industrialized countries. In some countries, indigenous sources of venture capital or commercial interests may be available.

In efforts to stay abreast of developments, several LDCs have set up national institutes or programs in biotechnology. Many rudimentary techniques, such as tissue culture, are being widely adopted. However, faced with mounting national debts and severe budget constraints, many countries are finding it increasingly difficult to maintain, let alone increase, the level of investment in agricultural research. Given budget constraints, developing countries need to determine which biotechnology areas should be given priority. Equally important, they need to know what training their scientists require to enable them to adapt biotechnologies for local use. In partial answer to these questions, experience shows that the first requirement is a strong program in the applied agricultural sciences involving plant breeding, plant protection, and agronomy. Specialized training in biotechnology technologies can then follow if resources permit.

Potential Role of the International Agricultural Research Centers

Many people anticipate that in the area of major food crops the International Agricultural Research Centers (IARCs) will play the same central role in biotechnology transfer that they did earlier in the Green Revolution. Others argue that direct links between National Agricultural Research Centers (NARCs) and developed-country institutions would be more efficient. Both approaches are likely to be used and will require the development of close institutional links to facilitate the two-way flow of information.

Such links are also needed to enable the developed-country laboratories to address the priority needs of the developing countries as these countries see them—not as they are seen by developed-country scientists. As one would expect, relatively little research is being done to develop biotechnologies for crops such as cassava, sorghum, millets, pigeon peas, cow peas, and plantain that are especially important to the developing countries.

One of the most important contributions of the IARCs is their participation in international networks which serve to link institutions and programs of mutual interest. As already mentioned, the Rockefeller Foundation supports a network on biotechnology for rice that links advanced laboratories in Europe, the U.S., and elsewhere with IRRI in improving rice production. The Advanced Cassava Research Network facilitated by the *Centro Internacional de Agricultura Tropical* (CIAT) is integrating biotechnology approaches in overcoming production constraints in cassava. CIMMYT recently initiated a network for the use of RFLPs (Restriction Fragment Length Polymorphism) in maize genome mapping.

The IARC's current involvement with biotechnology is sure to expand rapidly. A recent evaluation of international biotechnology efforts recommended that the IARCs consider undertaking these four initiatives (Plucknett, et al., 1989).

One: To classify agricultural research programs based upon their capacity to use various applications of biotechnology. This would assist the centers in determining the needs and abilities of their national program partners and ensure the transfer of high-priority technologies. In addition, it would help define outreach responsibilities associated with each new technology.

Two: To explore opportunities for establishing commercial relations so that revenues generated could provide an additional source of core funding to the centers. This would operate in a manner similar to that used by universities for the receipt of royalty payments. As noted previously, industrialized countries may become increasingly reluctant to allow donor agencies to provide support for center-based research on major commodities if it is perceived that this competes with their export capabilities.

Three: To establish Institutional Biosafety Committees (IBCs) to provide a responsible mechanism for the safe use and development of technologies and organisms in international research. Each IBC would coordinate testing with host-country approval mechanisms, as well as with current regulatory standards adopted by the developed countries. The composition of each IBC would vary from center to center but should include scientists and other representatives from the host country.

Four: To establish a long-standing expert group to deal with the role of biotechnology in world agriculture. This would need to be a standing group rather than a one-time meeting of experts. While each center is capable of determining its own course, active dialogue with a group of experts should help each institution make better decisions.

In summary, research on biotechnology offers the potential for increasing agricultural productivity and reducing environmental problems. This appears to be the case even in many of the regions of the developing world that did not benefit significantly from Green-Revolution technology. However, many of the developing countries, particularly those in Africa, do not have the capacity in research and extension required to adapt biotechnology to local conditions. A few Third World countries will be able to benefit from association with transnational private firms. Some countries with strong national research programs, such as India, Brazil, and China, may be able to work directly with research laboratories in the developed countries. But for the bulk of developing countries, the network of IARCs is likely to provide a key access link to biotechnology.

Most of the IARCs are already engaged in applied biotechnology research, and some have extensive links with research laboratories in the developed world. As they move into the age of biotechnology, the IARCs will be pressed to strengthen their capacity in order to (1) identify and transfer high-priority technologies (or end products developed from biotechnology) to developing countries; (2) explore opportunities for establishing commercial relations with private-sector companies (where this can be achieved without conflict of interest); (3) encourage biosafety; and (4) establish an expert group to guide biotechnology in world agriculture.

Initiatives for International Agricultural Research Centers

- *identify and transfer high-priority technologies to developing countries*
- *explore opportunities for establishing commercial relations with private industry*
- *establish institutional biosafety committees to coordinate the safe use and development of technologies in international research*
- *establish a standing group of experts to deal with the role of biotechnology in world agriculture*

Conclusions

The conclusions which follow relate to the issues and policy options that emerge in the four main sections of the chapter. For each of these, the issues that emerge from the chapter and the policy options are discussed.

Private- and Public-Sector Interests

The change in U.S. patent laws makes it possible for private firms to capture a major share of the profits from investments in biotechnology research. Already the private sector is playing a major role in the development of new biotechnologies. A new division of labor is developing between the public and private sector, with the former concentrating on basic research and the latter on technology development and dissemination.

Taken to the extreme, such a division of labor would likely limit the development of biotechnologies which have a high social but low private benefit. Furthermore, without strong public-sector capacity in biotechnology development, advances in biotechnology in the Third World would be limited to that handful of countries where private transnational firms would find it profitable to operate. Lack of access to successful biotechnology development would have a negative impact on income distribution at all levels—local, national, and international.

Both the private and public sectors should be encouraged to develop and diffuse biotechnologies. This may require that the public sector create new organizational forms of technology development and transfer. Already this has occurred with the Michigan Biotechnology Institute (whose focus is on biotechnologies important to the state of Michigan) and the international rice biotechnology network, supported by the Rockefeller Foundation, which links advanced laboratories in the developed countries with applied research stations in the developing countries.

The Competitiveness of U.S. Agriculture

There is a growing concern about the ability of U.S. agriculture to compete successfully in world markets. In part this concern is related to the fact that our competitors (in the developed and developing world) are taking advantage of U.S. advances in basic science and biotechnology to strengthen their own competitive positions. Many argue that access to such technology should be further restricted through stronger protection of intellectual property rights. In a somewhat different area, U.S. legislation already exists which prohibits foreign financial assistance for research on crops that would compete with U.S. exports.

Of concern here is both the size of the world export market and the U.S. share of that market. The U.S. has by far the largest research base for biotechnology of any country in the world. However, it is extremely difficult, even if it were desirable, to create a strong barrier of protection for intellectual property rights. This is probably fortunate. Short-run gains in U.S. market share as a result of such protectionism could be more than offset by a slowdown in world agricultural and economic development. In the longer run, this would result in a decline in demand for U.S. exports.

We conclude, therefore, that further strengthening of laws to protect intellectual property rights is likely to be of limited significance in maintaining American competitiveness in agriculture. As indicated by studies such as that of the Office of Technology Assessment, key determinants of U.S. competitiveness are financing and tax incentives for private firms; the level of government funding for basic and applied research; and personnel availability and training.

Concerns about health and environmental hazards will block or delay the acceptance of certain biotechnologies in the U.S. and other industrial countries. This in turn could adversely influence the competitive position of U.S. agriculture in those commodities affected.

Impact of World Trade

Biotechnology enhances the capacity of industrial countries to create substitutes for imports (from developing countries and elsewhere). The protectionist policies of the developed countries create further incentives for scientists to search for such substitutes.

Sweeteners are an example of the impact of such policies. Legislation designed to protect domestic sugar beet producers encouraged the development of the corn sweetener. This led to a sharp drop in the demand for cane sugar (which is produced largely in developing countries). How successfully and how rapidly other substitutes can be developed using biotechnology is difficult to predict.

In the current GATT negotiations, efforts are being made to encourage trade liberalization in agriculture. Such efforts should be coupled with greater attention to strengthening applied-research and technology-transfer capacity in the developing countries. Strengthening trade linkages between developed and developing countries through trade liberalization will work to everyone's advantage. Failure to achieve this will lead to a continuing slowdown in world trade and growing efforts worldwide to have self-sufficiency in agricultural commodities. In the long-term, such developments would be detrimental to U.S. agriculture and to the economies of Third World countries.

Technology Transfer to Developing Countries

Biotechnology is knowledge intensive. Most of the advances in basic research and technology development are likely to occur in the industrial countries. While the potential for increasing the productivity of developing-country agriculture is great, appropriate technologies tend to be very site specific, and many developing countries lack the capacity to adapt biotechnologies to local conditions. Further, most of these countries are not in a position to attract private investment to this area by transnational firms.

Although advances in biotechnology should lead to a general increase in agricultural productivity, the productivity gap between developed and developing countries is likely to widen. Furthermore, the productivity gap is likely to widen between the dozen or so most advanced developing countries and the rest, for the latter lack the basic institutions and infrastructure to take advantage of advances in biotechnology.

Steps such as the following should be taken to encourage wider application of biotechnology in the developing countries and to lessen the negative distributional impact. First, developing countries will need to strengthen research capacity in the traditional applied-research fields such as plant breeding. Second, in the industrial countries more public-sector funds should be made available for biotechnology research on commodities and problem areas that are important to the developing countries.

Finally, to facilitate technology transfer, direct telecommunication linkages will need to be expanded between researchers in developed country laboratories and their counterparts in developing countries. Advances in information and communication technology can greatly facilitate the two-way flow of information. Because the IARCs are likely to play a central role in the transfer of biotechnologies to developing countries, they will need to expand their capacity for assistance in this area.

Final Observations

Biotechnology appears to offer great potential for future productivity growth in agriculture. However, as with all technological change, the benefits are not likely to be spread evenly. Maintenance of U.S. competitiveness in world markets will depend principally on the level of government funding for basic and applied research.

However, it is extremely difficult to predict the impact of biotechnology on the volume of U.S. agricultural exports because this will depend on a number of other factors. Much will hinge on the degree to which policies encourage economic growth in the developing countries, for they are the greatest potential source of increased demand for U.S. agricultural exports.

Two very different scenarios can be envisioned, one pessimistic and the other optimistic with respect to the prospects for economic development and U.S. export demand. If existing protectionist policies continue, the Third World debt problem goes unresolved and too little attention is given to the development of agricultural research capacity in the Third World, the slowdown in economic growth and agricultural trade witnessed in the past few years will become a chronic problem.

In this environment, advances in biotechnology will tend to increase inequities, not only between developed and developing countries, but also among the more and less advanced developing countries. So those countries with the most highly developed agricultural research systems will gain relative to the rest.

On the other hand, if the structural maladjustments of the 1980s are reduced or removed and, through technical assistance and other measures, applied-agricultural research capacity is strengthened throughout the Third World, then the impact of advances in biotechnology is likely to be more evenly distributed. If productivity growth can be realized over a broader range of countries, the high export levels of the 1970s could become a reality again in the 1990s.

References and Additional Readings

Buttel, Frederick H., Martin Kenney, and Jack Kloppenberg, Jr. 1985. From green revolution to biorevolution: Some observations on the changing technological bases of economic transformation in the Third World. Economic Development and Cultural Change 34(1): 31-55.

Congress of the United States. 1989. Agricultural progress in the Third World and its effect on U.S. farm exports. Congressional Budget Office, Washington, DC.

Fransman, Martin. 1986. Technology and economic development. Wheatsheaf, Brighton, England.

Fransman, Martin. 1991. Biotechnology—generation, diffusion, and policy: An interpretative study. Mimeo, University of Edinburgh.

Plucknett, Donald, Joel I. Cohen, and Mary E. Horne. 1991. Future role of the IARCs in the application of biotechnology in developing countries. In *Agricultural biotechnology opportunities for international development. C·A·B International, U.K.*

Presley, Gabriel, ed. 1991. Agricultural biotechnological opportunities for international development. C·A·B International, U.K.

van den Doel, Kees and Gerd Junne. 1986. Product substitution through biotechnology: Impact on the Third World. Trends in Biotechnology.

CHAPTER NINE

Biological and physical scientists ordinarily have philosophic and methodological orientations that put investigation of ethical issues and questions beyond the realm of objective research. However, ethics, which is particularly concerned with procedures or rules for converting both value and value-free knowledge into prescriptions about "what ought or ought not to be done," has an important role to play in our public choices and private decisions about agricultural biotechnology.

Ethics requires that both value and value-free knowledge of biotechnology and its impact be developed and used in making public choices and private decisions to handle risks of harm and attain the benefits of biotechnological products and techniques. Biological and physical scientists can contribute by providing objective knowledge about possible biotechnological benefits and damages, including the probabilities of their occurrences. Social scientists and humanists are in a position to contribute by helping to develop knowledge about the monetary and nonmonetary badnesses of damages and the monetary and nonmonetary goodnesses of benefits from biotechnology.

While our discussion of values currently relies on rather abstract arguments involving such things as populism, agrarianism, nature, and different theological perceptions of the relation of God and man to nature, these abstractions will be supplemented with increased experiential knowledge as our experience with biotechnological processes and events increases. Those who represent agriculture and the public interest need objective value and value-free information that has been tested logically and experientially if they are to make the right decisions—and be effective in presenting them. In addition to the risk that we will do something we "ought not to do," agricultural and scientific leaders ill-informed about ethical issues will be unable to make an effective case for biotechnological breakthroughs in the public forum, causing us to fail to do what we "ought to do" with agricultural biotechnology.

9 Ethics and Values Associated with Agricultural Biotechnology

Glenn L. Johnson, Michigan State University
Paul B. Thompson, Texas A&M University

Introduction

Ethics is evaluative of choices, decisions, and acts, both public and private. In reaching evaluative conclusions as to what "ought or ought not to be done" or "what ought to or ought not to have been done" about different aspects of biotechnology, two broad kinds of information are required. One of these has to do with values as perceptions of the goodness and badness of conditions, situations, things, and, indeed, of acts themselves. The second concerns the value-free characteristics of conditions, situations, things, and acts. Biological and physical agricultural scientists specialize in producing the second kind of knowledge.

Because our knowledge of both the value-free and value characteristics of biotechnology and its consequences is very imperfect, learning and research (which reduce imperfections in our knowledge) are important aspects of the ethics of biotechnology. There is an ethical obligation to learn and become informed with respect to both kinds of knowledge before reaching conclusions about what ought or ought not to have been or to be done vis-à-vis biotechnology.

Much of what is discussed in this chapter as being directly related to biotechnology perhaps should be considered in the context of continuing developments in biological science. After all, biotechnology may be considered as a set of additional tools for influencing genetic change. In that context, biotechnology represents a process for influencing change rather than a change (or product) per se.

Prescriptions are generated by processing value-free and value knowledge through decision- and choice-making rules. Prescriptions differ from value knowledge. The word "prescription" is used here to stand for the rightness and wrongness of acts (past, present, and proposed)—with what ought or ought not to be or to have been prescribed and done. By contrast, values deal with goodness and badness.

The difference between prescriptions and values is comprehended when we realize that we sometimes ought to do an act with bad consequences if they are the least bad that can be done in the circumstances and, reciprocally, that we often ought not do an act with good consequences if an act with still better consequences can be done in the circumstances (Lewis, 1955). Thus, choices and decisions about biotechnological actions may be prescribed as right even if they have bad consequences and proscribed as wrong even if they have good consequences. The situation is further complicated if the act itself is good or bad regardless of its consequences.

Involved here are questions about what probabilities should be tolerated in our choice and decision rules of doing the wrong thing—of failing to do that which ought to be done or of doing that which ought not to have been done. Questions about social, religious, political, and police power are also involved. As we gain more experience in making choices and decisions about biotechnology, we find that we are often involved in establishing institutions that redistribute the ownership of rights and privileges and, hence, power among different groups.

These preliminary comments suggest some important ethical questions. What do we do when our knowledge of biotechnology is as incomplete, fragmentary, and inconclusive as it now is? Are we ethically obligated to improve our knowledge basis? How much are we obligated to learn before we draw conclusions about what ought or ought not to be done or what ought or ought not to have been done? What are our obligations to be unbiased and truthful in our communication with others about biotechnology? Do these obligations apply to those with vested interests (including scientists) in advancing or retarding biotechnological methods and techniques? Is it ever ethical to decide to harm some persons in order to benefit others? How do we take the welfare of others into account?

We first consider some of the above questions more specifically and concretely in discussing biotechnology's impacts on the environment, biotechnology as a form of playing God, and biotechnology's impacts on food safety and quality. We then take up the more abstract issues of taking risks for gain, the impacts of new techniques and products on farms and agribusinesses, and, finally, the pursuit of values that compete with attainment of other values. As we proceed through these discussions we will proceed from "lay person ethics" to somewhat more advanced ethical considerations. Our chapter ends with a summary and conclusion.

Impact of Biotechnology on the Environment

Advances in biotechnology are anticipated to have damaging as well as beneficial impacts on our environment. Opponents of biotechnology stress its possible damages, while proponents emphasize the benefits. Ethics requires that both be considered. Currently, very few of the possible environmental damages and benefits of agricultural biotechnology are well known. For many of those known to be possible, the probability of their actually occurring is not known.

Since biotechnology is anticipated to have damaging and beneficial impacts on our environment, ethics requires that both potential outcomes be considered.

Making ethical decisions about the environmental impacts of biotechnological advances requires us to ascertain what environmental consequences (both beneficial and harmful) are possible and to understand them. Scientists and agriculturalists are responsible for producing such knowledge, but this does not mean that they have done so. It takes time and resources to produce such knowledge and sometimes urging from other concerned groups.

In addition to knowing what kinds of impacts are possible and the probabilities of their occurring, we need to determine how much and what kinds of values are likely to be experienced. Value knowledge about damages and benefits will be better if those who will experience the benefits and damages are interactively involved with those who are investigating the goodness and badness of environmental impacts (Johnson, 1987a).

The phrase "environmental values" covers a wide variety of nonmonetary values associated with environmental changes such as the following.

The Values of Human Health and Safety: Chemicals and organisms produced through biotechnology that can cause illness and injury when released into the environment raise questions about whether the consequent harm will be offset by good and, in turn, how both are to be distributed and managed.

The Value of Natural Resources: Humans depend upon nature in subtle ways. If biotechnological products affect climate or diversity of species, for example, future access to the natural resources needed to support human life may be threatened.

Aesthetic and Cultural Values: Cultivated farmland, wild areas, and watersheds are valued for the beauty, serenity, and continuity they provide with the past. If biotechnological products and techniques change the appearance or makeup of rural areas, these values will be affected—perhaps adversely but also perhaps in a favorable manner.

The Value of Habitat: The natural environment is home for many plant and animal species. Adverse environmental impacts can make their survival more difficult. Though it is clear that the resource value of plants and animals to humanity should be reflected in public policy, it is controversial whether costs borne by and benefits conferred on non-humans "count."

The Value of Stewardship: Some people believe that acts to protect nature and to preserve environments in their natural state are valuable. They see doing avoidable harm to plants, animals, and natural systems as a failure to fulfill a moral duty.

It is important to recall that environmental values are among many values potentially affected by biotechnological products and techniques. Obtaining any of the values considered above typically entails monetary and nonmonetary sacrifices. Regulatory systems now being devised and proposed for biotechnological products and techniques should balance the environmental values listed here not only against their monetary and nonmonetary costs but also against each other.

A key question for environmental regulators of biotechnology is how to devise regulatory procedures when there is disagreement as to what values are at stake. A regulatory framework that ensures protection of wildlife habitat might compromise human use of natural resources and impose direct costs (in the form of taxes and restricted use) upon some individuals who do not recognize the value of reducing harm to nonhuman animals, plant communities, and ecosystems. To be successful in designing redistributive governmental regulations, scientists, citizens, and governmental officials must engage in an ongoing dialogue on the meaning of environmental values and on standards of fair play and justice (Arrow, 1963; Johnson, 1987a).

The words "nature" and "natural" indicate a broad range of values. People experience the goodness of natural beauty in a mountain range, a seashore, or a bluebird. Some prefer anything that is natural to artificial substitutes, without regard to differences and similarities in other attributes of the alternatives. Natural foods, natural ability, and naturally curly hair are often venerated. The term "human nature" is used to indicate the core characteristics and tendencies of our species. Saying that someone is good-natured is a form of praise.

Value statements in which variants of the word "nature" play key roles can be extended indefinitely. Though it is not always obvious why something that is natural should be good, philosophical analysis and examination of experiences can help indicate whether it is or is not good.

Many physical and biological scientists regard natural things and processes as neither intrinsically good nor bad. Natural processes allow organisms to reproduce, sustain themselves, and interact with a broader environment, but neither organisms nor processes have value in virtue of being natural.

Since biological processes are instruments, so the physical and biological science view says, there is no intrinsic value difference between biological entities that have evolved through natural selection versus those created in a laboratory. That is, there is no intrinsic difference between entities that originate in the process of natural cell mutation and those that originate in a process in which a scientist alters a section of a genetic code.

In both cases, the resulting biological entities are evaluated in terms of their ability to help us attain the values we seek at low cost. Thus, so long as biological processes are judged instrumentally, it is hard to see why nature and the natural should have values at all, yet it is clear that many people do experience the value of nature itself.

Logical attempts to defend or establish natural values on the basis of instrumental arguments are seldom successful, probably because natural values are *more experienced than defined.* For example, when one argues that the wilderness is just an instrument for producing certain feelings of solitude and harmony, it is the feelings that are valued rather than the natural character of wilderness.

Yet people almost uniformly report that they are more satisfied with a truly wilderness experience than with an artificial wilderness experience, even when they cannot detect differences between the artificial and natural setting. Apparently they experience the value of nature itself and its authenticity and integrity as ends, not as means. Such experiences indicate that nature has intrinsic as well as extrinsic or instrumental value.

These arguments point to some enduring philosophical issues that were touched on in the questions we posed at the beginning of this chapter. There is a deep-seated tension between approaches that lead us to (1) understand biological processes as instruments for the production of valued products and events and those that lead us to (2) identify from experience more about the value of nature and its associated elements of authenticity and integrity. It may be possible for us to accommodate these different views through research on and public discussion of the links between instrumental or exchange and intrinsic values.

Biotechnology as a Form of Playing God

From the earliest days of human history, the idea that human beings would one day unravel the mystery of biological creation and gain the power to manipulate life forms has raised deep religious questions. Mary Shelley's *Frankenstein* was an extended allegory about the tragedy and evil that were unleashed when an impious scientist overstepped the natural boundaries of human action and created life. Biotechnology does not create life from dead matter; instead, it increases our ability to control the forms of already living things. Nevertheless, since the first discovery of recombinant DNA, the rise of genetic engineering has led religious people to ponder its meaning.

Most religions espouse certain values and moral responsibilities as central to the faith but also hold that membership in the faith or belief in its central tenets is not a prerequisite for living a moral life. Theological ethicists typically feel that it is important to explain their key ideas in language that is traditional to the particular religion and denomination in which they work, while philosophical ethicists feel that it is important to explain their key ideas in language that makes no reference to any particular religious tradition. There are no ethical choices on which theological ethicists have universally (or even substantially) taken a unified position against that taken by philosophical ethicists, but religious ethics does introduce an additional source of reasons for expressing and understanding goodness and badness and for distinguishing right from wrong.

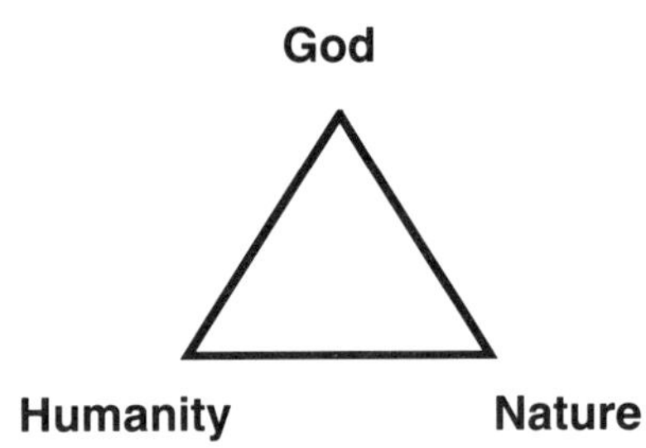

Religious ethics sometimes describes moral duties to do the right thing as duties to God, rather than as duties to other people. For example, the ethical principle that it is usually wrong to kill innocent people is recognized almost without regard to religious creed. One way to understand this principle is to think that it would be wrong in virtue of unjust harm done to the innocent person. Another understanding comes from thinking that it is wrong because killing violates one of God's commandments. In the second case, the wrong is conceived to be done to God, not to the innocent victim. Those who accept the principle can live together in a civilized society, regardless of why they think killing is wrong.

The question here is whether the manipulation of recombinant DNA to modify plants and animals and to improve production processes for agriculture violates basic principles of religious ethics. Philosophical and religious ethics coincide on much of what has been stated above. Some people, however, experience a feeling of religious dread when confronted with biotechnology, and some political opponents of biotechnology have found a great deal of support among conservative Christians. One reason for this may be the fact that tissue from aborted human fetuses has been associated with some recombinant DNA research, and groups on record as supporting the fetus' right to life have extended their concern to opposition of such research. This argument does not address biotechnology directly, however, but only one of the many research techniques used in conjunction with recombinant DNA research. Of more concern is the opinion that biotechnology is a kind of Frankenstein, a way of playing God by failing to respect human limitations.

The idea of God and of God's commandments is a key to understanding Christian ethics, but other ideas are also important. One is the idea of nature. The interpretation of nature has a mixed history in Judeo-Christian thought. Some have thought that God allows humanity free reign over nature and that human use of nature cannot have moral significance. In this case, nature is viewed as a stage on which the human drama is played out; the stage itself has no intrinsic value and, indeed, must be spurned in pursuit of heaven.

Others have thought of nature as God's masterwork and see the human race as only a component of God's creation. According to this view, even if humanity is the crown of God's creation, the crown takes on its meaning within the context of God's greater whole. The former view has sometimes encouraged people to place no value on plants, animals, and the wider ecosystem apart from the wealth that can be extracted from them, while the latter view has sometimes encouraged them to see nature as sacred and to see acts that affect natural processes as subject to religious constraints.

Sometimes the interpretation of nature refers to the interpretation of divine creation. Judeo-Christians place great significance on God's power of creation. Those who see nature as humanity's dominion believe that plants, animals, and ecosystems were created by God for human use. Those who see nature as sacred believe that creation is a divine activity that must be respected when encountered in nature.

It is possible to interpret this "respect" as prohibiting biotechnological research, but whatever one believes about God, human beings do have considerable creative powers of their own, presumably God given. Belief in the sanctity of God's creative powers must recognize that human beings were created to be creators, too. Thus, there must be at least some creative human acts that are consistent with respect for the sanctity of God's creative power. The question is: Does the new biotechnology fall within the sphere of human creativity or does it infringe upon divine creativity?

Recombining genetic material may be just another use of natural materials for those who see nature as humanity's dominion. For those who see nature as sacred, the problem of finding limits to human action is more important. Ervin Chargoff (1976), a formidable scientist who led research on recombinant DNA, felt that "evolutionary wisdom" represents a kind of sacred border that humanity should not cross.

Some of those who have argued that biotechnology poses a threat to natural diversity and in turn to the chance interbreeding of races and species have invested this concern with an almost religious commitment to natural creativity. Here, religious concern begins to refer back to the environmental values we discussed earlier. This issue cannot be settled in this context, but it is one that promises to reappear at intervals in the research and development of agricultural biotechnology.

Food Safety and Quality

Abstractly, assessing biotechnological results is a little different for food-related than for environment-related products and techniques. However, assessment of food-related products and techniques does illustrate nicely some of the ethical questions considered in the introduction to this chapter.

In making decisions about which biotechnological food products we will consume and how much of them to use, we will often have to trade values off against each other; that is, we will often have to give up attainment of one or more values in order to obtain others. The values that we sacrifice in making a trade-off are sometimes more intrinsic to food safety and quality than the exchange values we establish in making trade-offs.

For instance, in the future we may have a new biotechnological apple variety that is very cheap to produce. In such an instance we might be willing to give up some taste or crispness in order to benefit from a large quantity of lower priced apples. In trading off taste and crispness for lower costs, we may even discover that the new apple has other characteristics of value. If the new apples produce more juice, their crispness will be less important while their taste may be enhanced with blends of other juices.

Though the exchange values we generate in balancing values such as

Giving up some taste or crispness in apples in order to benefit from a large quantity of lower priced apples is one of the trade-offs we might decide to accept.

We should not wait to start the research and education efforts to generate and distribute value knowledge until after biotechnological research produces its results.

taste and crispness against costs may ultimately be reflected in market prices, not all trade-offs generate monetized values. For instance, we might decide politically that we should not increase exposure to a food carcinogen in exchange for a lower priced or otherwise more convenient biotechnological product. In this case, we would probably decide that the goodness of the money saved is not worth the badness of a significantly greater carcinogenic exposure.

The exchange values generated in political, social, religious, and market processes are affected by who holds power within a society and by rules and regulations on exchange processes. For example, power to affect market prices depends upon having income to spend. Exchange values are also affected by who holds social, political, religious, police, and military powers. They also depend on having knowledge about one's choices. When distributions of power are changed, exchange values change even though intrinsic values may be unaffected.

Some of the ethical issues that are arising in connection with biotechnological advances involve changing the distributions of power in the "rules of the game" (Bromley, 1989). Sometimes such changes are themselves expensive in terms of human effort and goodwill, as well as money.

Persons and groups concerned about the effects of biotechnology on food safety and quality often want to change the locus of power for making such decisions. The exchange values generated by new biotechnological products will be vastly different depending upon whether decisions about which products to develop are made by commercial firms hoping to profit from new products or by civil servants and academicians who expect that R&D costs will not be recouped in the product's market price.

Land-grant universities, for example, have traditionally produced and made new seed varieties available at prices which do not cover development costs, much less allow for profit. If the roles and decision rules are changed so that private firms may recoup investment costs and profit from new varieties, exchange values will be affected. Some development costs will be shifted from tax revenues to market prices for farm inputs. These changes will, in turn, affect who produces food, who consumes it, and, finally, whether public universities even have the capacity to do research in biotechnology.

How are food quality and safety affected by such complex changes? Not enough is known about the answer to this question. Biologists, other researchers, farmers, agribusiness leaders, and activists need to know and understand the costs and impacts of changing power structures, roles, and rules of the game. This is in addition to the better knowledge they need of the more intrinsic "real" values of biotechnological products and methods.

We should not wait to start the research and education efforts to generate and distribute value knowledge until after biotechnological research produces its results. Such knowledge is needed to help establish priorities among the different kinds of biotechnological results. The range of outputs that can be attained through research in biotechnology promises to be so large that we should devote our biotechnological research resources mainly to those we anticipate will be useful and used.

Agriculturalists should also know that there is important, well-recognized, scholarly literature on ethical aspects of making public choices about issues and problems such as those about biotechnology. It includes work by Oliver Williamson (1985) and James Buchanan and Gordon Tullock (1962), two economists and a legalist; Douglass North (1987), an historian; and Daniel Bromley (1989), an agricultural economist. James Buchanan received a recent

Nobel Prize in economics for his work in this area. Earlier, Kenneth Arrow received a Nobel Prize for his research on social choices and individual preferences.

Agriculturalists with deeper, more advanced interests in these matters may want to consult some of the references at the end of this chapter. Others can derive comfort from the existence of this body of peer-reviewed, well-recognized literature.

This literature deals with the costs (nonmonetary as well as monetary) of redistributing power and changing the rules of the game. It also deals directly or indirectly with the costs of dismantling old and establishing new power structures and rules. Such costs are called "transactions costs."

Though transactions costs are encountered in major confrontations such as revolutions and wars, they are also involved when we try to change power structures or regulatory rules to control the impacts of biotechnology on food safety and quality. Both proponents and opponents of biotechnology need to know and understand the costs and impacts of changing power structures and the rules of the game, including the effects of such changes on prices and nonmonetary exchange values. This is in addition to the better knowledge they need of the more intrinsic "real" values of biotechnology methods and the products produced through biotechnology.

It must be recognized that research on values and public choice is inherently more difficult than research in biotechnology itself. This, however, neither detracts from its importance nor indicates it should not be done.

Research attempting to describe "real" values requires that the proposition that such value research is unobjective and unscientific be rejected (Kaplan, 1968; Achinstein and Barker, 1969). Social scientists have long recorded and analyzed monetary exchange values. They know and understand a great deal about how market, political, and social forces generate changes in prices, incomes, and expenditures.

Fortunately, objective research on exchange values need not be and has not been confined to monetary values. There is now the established, scholarly *Journal of Agriculture and Human Values* regularly publishing peer-reviewed research on values in agriculture. It includes papers on what can be labeled "biotechnological ethics." Consumption economics and welfare economics deal with the establishment of non-monetary exchange values. Risk assessment work, both theoretical and empirical, maximizes expected nonmonetary values (Friedman and Savage, 1948; Tversky and Kahneman, 1981).

More scholarly research is needed on nonmonetary values, and progress is being made. Biotechnological ethics, like biotechnology itself, is a growing field of academic endeavor.

Regulating, Insuring Against, and Taking Risks for Gains

True or not, it is rather widely perceived that the procedures and products of biotechnology will generate important dangers for individuals, various groups, and the environment. Research can help establish the nature and probabilities of adverse events and their badnesses, but if research is to improve the accuracy of our perceptions, we will have to devote considerable energy to including concerned parties in the research process. As knowledge of biotechnological events improves, it will become possible for individuals and society to devise ways to avoid unfavorable while seeking favorable events.

For unfavorable events, we commonly devise ways both to *insure* and *ensure*. When we insure we devise ways to compensate for the harm that occurs when adverse events take place. Insurance works well when the probabilities of adverse events are well known, when events are beyond the control of both the insured and insurer, and when the insurer's knowledge of values is good enough to know the badness of the premium and unfavorable events.

When favorable events are also possible, their probabilities and goodness also need to be taken into account. A difficulty with insuring (and also ensuring) against unfavorable biotechnological results is that we know so little about them—their nature, probabilities of occurring, and their badness. Such ignorance often breeds fear. Ignorance and fear combine to make it difficult for insurers to design acceptable insurance contracts and for individuals, firms, and organizations to design self-insurance strategies to protect them from unfavorable events caused by using the tools of biotechnology.

An alternative to insuring is ensuring—"to make safe or certain, to secure"—by controlling the probabilities that unfavorable events will take place. Proposals to regulate biotechnological research and the use of its products are examples of ensuring schemes. The famous Delaney Clause is an example of an ensuring strategy designed to virtually eliminate any probability of approving the use of a carcinogen.

Biotechnological ensuring has costs—for regulatory and monitoring activities and through elimination of possible favorable biotechnological events. Ensuring schemes work well when the following are well known: the reduction in probabilities of unfavorable events that results from ensuring, the nature and badness of the unfavorable events, and the costs of ensuring. Unfortunately, biotechnology is so new that such knowledge is poorly developed. Again, ignorance feeds fear.

We also have imperfect knowledge of the probabilities of favorable biotechnological results, some of which are so valuable that it is worthwhile taking "long shots" and risking substantial probabilities of unfavorable events to obtain them. Consider, for instance, the value of a possible AIDS vaccine that biotechnology may be able to generate in the next few years (Bolognesi, 1989).

What we know about chance taking or gambling strategies for pursuing important desirable events roughly parallels our knowledge of insurance strategies. We do best when we can identify the favorable events, know the probabilities of their occurring, and know their value or goodness. When such knowledge is poor, it is difficult to decide whether to go ahead with a particular biotechnology project. Something akin to pathological gambling can take place in the pursuit of favorable biotechnological results by research agencies when the values of favorable outcomes are high and the probabilities of their occurring are small.

When both important favorable events may be generated by a biotechnology project and when ensuring and/or insuring against unfavorable events are also involved, the picture is greatly complicated. In general, insuring and chance-taking theories are better developed than ensuring theory (Tversky and Kahneman, 1981). Theories for combining all three are as yet poorly organized (Johnson, 1987b, 1988). Thus, despite considerable empirical and theoretical work, the ethics of insuring, ensuring, and chance taking are not well understood or widely appreciated by administrators of the agricultural establishment.

Both ensuring and insuring become practical impossibilities under certain circumstances. One may ensure that one will not die from cancer as a result of smoking cigarettes by refusing to smoke cigarettes, but one cannot ensure that one will not die from cancer. Similarly, one may insure against monetary losses from theft of one's property, but one cannot insure against the emotional loss when the stolen item has sentimental value. Insuring is limited to losses that can be distributed or shared by a group of risk bearers. Ensuring is limited by capacity to forestall the unwanted event.

Ethics enters the picture because both ensuring and insuring involve tradeoffs. When a person ensures that he will not die in a subway accident by choosing to drive a car instead, he makes a poor trade, if the end is not to die in an accident. By analogy, a society that chooses to ensure against the low risks associated with consuming food products made available through biotechnology may "irrationally" choose an alternative with greater relative costs.

The most extreme strategy would be to ensure against harm arising from biotechnology by banning it entirely. Doing so, however, would mean continued health risk from present chemicals and from other undesirable situations and product traits produced by conventional means. It would also eliminate the desirable products research in biotechnology can produce.

Ensuring against biotechnological risk by eliminating biotechnology is probably a very ineffective way of averting the unwanted and largely unknown consequences we are actually trying to avoid. Nevertheless, some opponents of biotechnology advocate this form of ensuring (Rifkin, 1983, 1985).

Insuring offers different strategies. One form of insuring (self-insuring) is the limit case of playing the probabilities—recognizing that some unwanted outcomes will occur and simply accepting them as a "cost" of using the technology. A gambler knows that he will not win every hand and accepts the losing hands as part of the cost of gambling. Some have argued that society should be willing to accept the risks of agricultural biotechnology as a simple cost (Brill, 1985, 1988; Davis, 1987).

Another form of insuring is one that attempts to compensate harms incurred by individuals by distributing the total costs of compensation to all risk bearers and to those who receive benefits. Farmers who go bankrupt as a result of decisions made regarding new production techniques derived from biotechnology (or from new techniques developed by any means?) might be given compensation in the form of retraining or relocation assistance.

The ethics of risk distribution are not simple, however. In the pure case of a financial loss, the total value of all monetary losses can literally be distributed evenly across all parties to the insurance contract. As Douglas MacLean (1988) has shown, however, few pure cases of financial risk exist, and the distribution of personal risk is a tricky business. Although those who suffer adverse health impacts, for example, can be compensated, it is they and they alone who bear the physical and emotional consequences of illness.

Insuring also tends to place people in a passive role from which risks appear fixed. There are also cases, however, where the risks can be altered through ensuring into a more attractive alternative. Although changing one's options is an act that itself entails risks and costs, there are many cases in which ensuring can be accomplished at reasonable cost. Buckling one's seat belt is an example.

The problem of overlooking the importance of opportunities to ensure often occurs when comparing risks of two or more activities. One might think that anyone who rides a motorcycle would be willing to drive a car without a seat belt, since the statistical risk of motorcycle injury exceeds that of driving without a seat belt. But the relative costs of ensuring are totally different for the two activities, and the comparison is invalid. Some defenders of agricultural biotechnology advocate risk strategies that require calculation of probabilities but ignore opportunities for ensuring.

Ensuring and insuring are not mutually exclusive, of course, but risk issues become very contentious when any party to the debate insists on unilateral adherence to one approach or another. Food safety debates, for example, have been caught between the Delaney Clause, which requires ensuring against foodborne cancer risks without regard to other costs and benefits, and those who argue that acceptable levels of carcinogenic risks are determined at least in part by the costs of ensuring.

Biotechnology's most active opponents have insisted that the simplest ways to ensure against unwanted outcomes is to eliminate biotechnology or to slow the development of products and techniques so much that they become unprofitable to license and market. On the other extreme, biotechnology's proponents have argued that opportunities to ensure against unwanted outcomes can be ignored, since biotechnologies are already as safe as other techniques accepted by society. Ethics is concerned with finding an appropriate balance among ensuring and insuring against losses and taking chances for benefits.

Impacts of Biotechnology on Family Farms and Agribusiness

The products of biotechnological research can be expected to have many different impacts on agricultural production and agribusinesses. Biotechnology may lead to products that can be expected to replace chemicals and plant and animal varieties now in use. Persons employed in the production and distribution of some currently used inputs will have to adjust. Products from biotechnology may change economies of scale in agricultural production, making it impossible for some farmers who currently produce at peak efficiency to compete with larger operators who can capture the new efficiencies of scale.

Buckling a seat belt is an example of a case in which ensuring against a loss can be accomplished at reasonable cost.

In other cases, smaller part-time farmers may be the gainers. In either case, the consuming public "wins" in that new products and techniques are only adopted if they lower price relative to quality. Since consumers outnumber producers, many applaud progress in production techniques as "producing the greater good for the greatest number of people" (Madden and Thompson, 1987).

Generally speaking, much of what is held to be true about the impacts of unknown biotechnological products and techniques on farmers, agribusinesses, and rural societies is speculative and in need of further investigation, but some people will become "losers" who either must quit business or absorb costs. Is it fair that these people lose so that others may gain? The question is particularly relevant when the new inputs and methods have been developed by public institutions, using public funds.

Since family farms potentially can be either adversely affected or benefitted by agricultural biotechnology, society needs to examine the reasons for acting to ensure family farm survival.

There are two levels to the question as it affects the evaluation of agricultural biotechnology projects, products, and methods. First, it can be argued that fairness requires that those who are harmed when public agencies act to increase general welfare are entitled to some form of compensation. Second, some people believe that family farms have a special status that entitles them to a greater consideration than given to other businesses affected by changing production methods.

The fairness and compensation for harm argument applies equally to all persons adversely affected by the development and dissemination of all new production methods, not just those arising from biotechnology. Just as laborers who harvested tomatoes by hand may have been replaced by the mechanical tomato harvester, some people who work both on and off the farm may be displaced when new biotechnologies come on-line.

If society does not compensate the firms and workers that are displaced, they are forced to absorb some or all of the costs of liquidating investments and of retraining, relocating, and finding new ways to make a living. Decisions about whether they "ought to be" compensated are fundamentally ethical in nature.

Experience teaches that whether or not society has a moral obligation to compensate displaced firms and workers varies from one instance to another. A well-capitalized firm that fails to foresee the need to make changes may deserve to fail, but sometimes hardworking, productive individuals can be forced into poverty by changes that they could not be expected to have anticipated.

One rule of thumb is that society should make a particular effort to compensate losers from already disadvantaged classes (Rawls, 1971). A second is that those harmed by innovations originating in the public sector are entitled to public compensation, even when comparable harms resulting from private sector innovation would not be compensated (Nozick, 1974). As social scientists improve their ability to forecast dislocations, ignorance becomes an ethically less adequate excuse for poor public policy and planning than in the past.

Small family farms can be either adversely affected or benefitted by biotechnology. For instance, one of the most celebrated products of biotechnological research, synthetically manufactured bovine somatotropin (BST), has run into trouble because of alleged adverse effects upon small dairy producers. Some think BST may affect economies of scale in dairy production in a way that will make many small dairy farms unable to compete with larger ones.

Although the conclusion is controversial, the BST case is instructive because concerns about family farms have come to the fore. Family farms are sometimes thought to be entitled to protection from market forces that would cause their demise. This entitlement could mean that new techniques which reduce a family farmer's ability to compete should not be developed at all, however important the benefits they produce for others. Far beyond a one-time compensation for damages, family farms are sometimes thought to be entitled to continuing compensation to allow them to remain in operation, a cost that would significantly reduce the net value of a biotechnological product such as BST (Comstock, 1987).

In light of the seriousness with which harm to family farmers is viewed, it is worth examining some of the reasons that are given for ensuring family farm survival. These reasons are of three kinds. First, public institutions are thought to have contractual obligations to small family farms that must be honored. Second, family farms are thought to be particularly important for the survival of democratic institutions. Third, family farms are thought to be morally important because of their effect upon human character. Each of these is examined in turn below.

It is possible to interpret the 1887 *Hatch Act as a social contract* that established public-sector agricultural research under the aegis of the U.S. Department of Agriculture and the state experiment stations specifically to aid small farms. The act does not mandate a general goal of increasing farm productivity and omits direct reference to relatively large-scale agricultural producers, though it does stipulate that ensuring a safe and wholesome food supply is among its goals.

A relatively recent legal case that first went against the University of California before being overruled concerned development of the mechanical tomato harvester. The case was based upon language in the original Hatch Act. The plaintiffs argued that the act created a statutory obligation for USDA/land-grant research institutions to at least examine the impact of their research on small farms.

A popular argument, ascribed to Thomas Jefferson, ties the value of the family farm to populist political philosophy.

This interpretation can be viewed as implying an ethical promise to use public funds to benefit small farms, as well as to pursue other mandated goals. Such a promise would also mean that pursuit of other goals should not overwhelm research to benefit small farms and, therefore, that the USDA/land-grant system should not support research that will drive small farmers out of business.

On the other hand, it is not clear that current USDA/land-grant programs should be bound to a promise made more than a century ago. Nevertheless, the U.S. Congress has repeatedly renewed its commitment to small farms, and in the absence of investigations that could clarify the goals of agricultural research, the 1887 act gives small-farm advocates an argument against biotechnological products such as BST.

A second argument ties the value of the family farm to *populist political philosophy*, sometimes putting words into the mouth of Thomas Jefferson. Jefferson did argue that the predominance of small farmers among the U.S. population would help the new nation forestall the problems that followed the establishment of democracy in France. The excesses of mob rule and the capture of policy by self-seeking supplicants were, in Jefferson's view, symptomatic of urban industrial society. Farmers, he felt, were tied to the land and would identify their own interests with those of the state: protection of property and economic stability.

This Jeffersonian argument has been at least partially used by authors such as Harold Briemeyer (1965) and Jim Hightower (1978). They argue that small farms are important to democracy because they provide an opportunity for relatively unskilled working class people to enter into entrepreneurship so as to control their own destiny and be their own bosses. A capitalist democracy in which everyone in the working class has no option but to accept wage labor at the prevailing rate is vulnerable to criticism, but one in which working class people have meaningful opportunities for entrepreneurship (of which a small, start-up farm is one) can truly be said to be a society in which people have consented to lead the lives that they do.

A final area of concern has been labeled *"agrarian fundamentalism"* and the "agrarian myth" by those who wish to ridicule it. It is the idea that living on a farm—working with nature, respecting nature's cycles, and becoming proficient in the diversified tasks of general farming—leads to a well-balanced moral character. Ralph Waldo Emerson remarked on the virtues of farming by comparing farmers to city folk who, though "talkative and entertaining," are "artificial." Emerson was the most prominent of many nineteenth-century writers who thought that the increasing industrialization of American society would bring about moral decay.

This vision of morality stresses the formation of values and character traits in the conduct of one's total life. Family farms, it is held, are good because the family is brought together within a meaningful context of work; family members come to know and appreciate one another, and, as they do, their own roles and identities become clear. By contrast, it is held that families in which work serves only to bring home a paycheck have less integration and serve only as a consumption unit after children are of school age (Berry, 1977).

The idea that family farm members learn to be both self-reliant and dependent on one another is viewed as a natural result of the productive life of the farm. The practical requirements of farm life are thought to help people appreciate the value of work, of independence, and of ownership, and also to make it clear that success depends upon goodwill, cooperation, and fulfilling one's responsibility to others. Nineteenth-century Americans believed that a farm family could not be successful without acquiring these values but that agrarian values would be impediments to success for factory workers, merchants, and manufacturers.

Those who hold this belief today experience the transformation from agrarian to industrial society with a great deal of regret. In this view, any change that makes agriculture more industrialized and more subject to the type of management decisions that typify nonfarm businesses is a change that penalizes the virtuous in order to reward the sinful. These agrarian beliefs appear, at least, to provide moral grounds for resistance to changes in farm structure that threaten family farms. The accuracy of these beliefs is open to numerous questions.

Current farm management research indicates that research in agricultural biotechnology will indeed have impacts upon small farms, though much more study is needed to predict these impacts with confidence, what the impacts will be, and whether they will be favorable or unfavorable to small farms. Small farms are subject to a variety of forces that affect their financial stability and their character as a work place. The influence of inflation, exchange rates, wage rates, and commodity policies may so greatly outweigh that of technical innovation as to make biotechnology's impacts upon small farms relatively negligible.

In the absence of more good farm management, sociological, and economic research, it is very difficult for defenders of biotechnology to make informed responses to biotechnology's opponents who exploit fears that new biotechnological products and techniques will cause the ultimate demise of family farming.

There are few analyses that examine the link between the populist or agrarian visions of the value of family farms and the actual impacts that agricultural biotechnological products and techniques can be expected to have on farms of different sizes. Populist and agrarian ideals may already be dead issues among existing family farmers and, hence, may have few implications for biotechnology. Despite this, populist and agrarian values may still be important as political symbols that define America's national identity and culture, but it is not clear what relevance these values have for the development and implementation of biotechnological advances.

At present, it is primarily the opponents of biotechnology that argue populist and agrarian issues. Not surprisingly, activists exploit the symbolic importance of the family farm rather than provide objective analyses and descriptions of its content and implications (Harbaugh, et al., 1987; Ostendorf, 1987). The public interest would be served by a full airing of opinion of these issues and their relationship to biotechnological advances.

The Pursuit of Biotechnological Values that Compete with Other Values

A further area of broad ethical concern arises from the fact that when resources are used to develop agricultural biotechnology, other things cannot be done, since resources can only be put to one use at a time. This involves the notion of opportunity cost. Decisions that ignore opportunities to use resources in more productive ways are both uneconomical and ethically indefensible—they are wrong even if good is being done because something better still can be done.

Although it is not a serious ethical offense when an individual fails to optimize the use of his or her resources in some of the less important decisions of life, the commitment of large portions of our society's agricultural research capacity to agricultural biotechnology is no trivial decision. We must be satisfied that we do not misspend key resources of society on biotechnology when we forego these other opportunities for using those resources.

On first appraisal, U.S., Canadian, and European agricultural surpluses appear to raise questions about the appropriateness of additional research on output-increasing agricultural biotechnology. However, prospective growth in world populations, increases in per capita incomes in large areas of the world, and threats to world land and water resources indicate the wisdom of continuing to invest in technological improvements for agriculture (Johnson and Wittwer, 1984).

Driving Forces for Farm & Rural Development

- *production efficiency*
- *institutional improvements*
- *human development*
- *preservation of natural resources*

Despite this conclusion, ethical questions must still be raised concerning our neglect of the three other driving forces for farm and rural development, those three additional forces being institutional improvements, human development (sometimes called human capital formation), and greater availability of natural and manmade resources.

Increasingly, the four driving forces appear to be so interdependent that imbalance among them means that any one of them present in larger than needed proportions contributes very little to farm and rural development. Domestically, U.S. rural communities are populated by disadvantaged and inadequately served rural farm and nonfarm people. U.S. population dynamics now indicate that an alarming proportion of entrants into the national labor force 10 to 20 years hence will be poorly trained and educated descendants of today's disadvantaged poverty households, many of which are female headed and in rural areas.

The quality and quantity of our natural farm resources are threatened, and there is much evidence of environmental pollution and contamination of environmental food chains. Rural family structures are threatened by drug abuse, teenage pregnancies, and social instability. The U.S. is neglecting human development, institutional improvements, and natural and manmade resources. We have an ethical obligation to bring our agricultural organizations (including land-grant universities) and governmental agencies into better balance. The agricultural establishment should not persist in its present neglect of our rural institutions, the development of our rural people, and the development, enhancement, and preservation of our rural resources.

There are two ways to get the resources required for the other three driving forces: (1) divert resources presently used for technological development to institutional, human, and resource development or (2) obtain additional resources for these other driving forces. If biotechnology is as effective as some of its advocates contend, then fewer resources for production research concentrated on biotechnology should permit us to maintain or even increase our rate of technical advance while devoting more resources to the other three driving forces.

However, this conclusion seems doubtful because a high proportion of our present technical advances in agriculture comes from nonbiotechnological research involving chemistry, physics, traditional plant and animal breeding, and engineering. Some agricultural research resources for generating technology could probably be saved through careful culling of both biotechnological and nonbiotechnological projects. This would concentrate our efforts to improve agricultural production on the most cost-effective projects in both groups. However, obtaining the needed additional resources for the three neglected areas from other sources appears more promising than reallocation from present physical and biological production research for agriculture.

Serious ethical questions will continue to exist until agricultural decision makers resolve issues about relative emphasis on the four driving forces and the emphasis on biotechnology within efforts to assist production agriculture. There are no clear answers now, but there is a clear ethical need to address these allocative questions.

Summary and Conclusion

As ethics is evaluative of public and private choices, decisions, and actions, it has an important role to play in our choices, decisions, and actions vis-à-vis biotechnology. Agricultural biotechnology promises many benefits for producers, consumers, and citizens. There are also possible unfavorable impacts.

Agricultural biotechnology is so new that many of its possible benefits and damages are presently unknown. Even for known or expected benefits and damages, the probabilities of their occurrences are poorly known. Still less is known about the goodnesses and badnesses of biotechnology's possible benefits and damages. Ignorance breeds fear of unknown damages by biotechnology's opponents, while magnification of possible benefits probably begets some undue optimism and unjustified taking of chances by its proponents.

Biological and physical scientists can contribute to biotechnological ethics by providing objective knowledge about possible benefits and damages resulting from the use of new biotechnologies, including the probabilities of their occurrences.

Social scientists and humanists are in position to contribute by helping to develop knowledge about the monetary and nonmonetary badnesses of damages and the monetary and nonmonetary goodnesses of benefits from biotechnology. Ethics (which is closely related to economics) requires that both value and value-free knowledge be developed and used in making public choices and private decisions to handle risks of damage and to attain the benefits of biotechnology.

Ethics is particularly concerned with procedures or rules for converting value and value-free knowledge into prescriptions about "what ought or ought not to be or to have been done." Voting, market, police, social, and religious power distributions affect decisions and choices. Paradoxically, progress in biotechnology changes power distributions, including distribution of the power that will come from creating new knowledge through biotechnology.

Biological and physical agriculturalists ordinarily have fundamental philosophic and methodological orientations that put investigation of ethical issues and questions beyond the realm of objective research. In this chapter, we have disagreed fundamentally with this orientation. Thus, we have had the difficult task throughout this chapter of dealing with such orientations so as to establish a legitimate place for objective logical and empirical investigations of the ethical aspects of biotechnology.

Our above examination of a number of biotechnological situations indicates that we have substantial bodies of logic and are gaining important descriptive value and value-free knowledge from our increasing body of experiences with biotechnological products and techniques. As our knowledge increases, fear (bred of ignorance) and undue optimism (unbridled by knowledge from experience) should both decrease.

In our discussion of various biotechnological issues, we considered ensuring and insuring as alternative ways of handling risks of future damages. Ensuring strategies attempt to reduce probabilities of damaging events taking place. Insuring leaves the probabilities alone but averages out the costs of damage among risk bearers.

Sometimes mixtures of insuring and ensuring strategies are used. The situation becomes still more complicated when desirable, sought-after events are in the picture, as they certainly are for biotechnological efforts.

Currently, the nature of future possible biotechnological events and the probabilities of their occurring are so poorly known that biotechnology's opponents can exploit ignorance and fears and "justify" unduly expensive ensuring schemes while, reciprocally, proponents may be making overly optimistic projections of favorable results in pursuit of funds for biotechnological efforts. Our general impression is that society at large is overemphasizing the dangers of biotechnology with too much ensuring. Time will likely confirm or deny this impression. Ethics requires the development of better, more objective value and value-free information on biotechnology to use in avoiding dangerous events while seeking the favorable products and techniques made possible through biotechnology.

We described various nonmonetary values in the previous section. We have also stressed that decision makers are ethically required to take knowledge of values into account in resolving biotechnological problems. Our discussions of values often relate the values that now influence choices, decisions, and attitudes about biotechnology to rather abstract arguments involving such things as populism, agrarianism, nature, and different theological perceptions of the relation of God and man to nature.

As our experience with biotechnological processes and events increases, these abstractions will be supplemented with increased experiential knowledge (both of values and value-free). This will increase our ability to make better informed and hence more ethical choices and decisions about biotechnological processes and results.

Honesty is an important part of ethics. Decisions are ethically questionable if based on inadequate amounts of knowledge, on knowledge that has not been adequately tested against logic and experience, on knowledge selected to support a vested interest while opposing knowledge is suppressed, or on falsified knowledge.

When values are not taken into specific account by agricultural decision and choice makers, it is likely that values untested by logic and experience will be used by activists who oppose agricultural interests in biotechnology. Conversely, proponents more interested in biotechnology than in helping farmers, consumers, and citizens can exploit untested values to avoid questioning of their work and its levels of financial support. Both opponents and proponents can find it advantageous to stress unobjectively held values that are of little direct concern to them.

Those who represent agriculture and the public interest need objective value and value-free information about biotechnology's impacts that has been tested logically and experientially if they are to make the right decisions about biotechnology and be effective in presenting the reasons for their decisions in political debate.

At present, there has been little systematic research on the nonmonetary values affected by agricultural biotechnology and even less research on how these values should be accommodated in biotechnological research and development. There is little evidence that public spokespersons for agricultural biotechnology are familiar with what is known about ethical issues. Activists opposing biotechnology, however, stress possible adverse impacts that agricultural biotechnology may have and make prominent reference to them in public campaigns intended to curtail research and development of biotechnological products and methods. There is a real danger that (1) institutional decision makers will neglect important nonmonetary values in setting policy on biotechnology and (2) spokespersons for agriculture will make ineffective responses to criticism of biotechnology advanced in public debate.

This danger may or may not have an impact upon the future of agricultural biotechnology. At present, decision makers within the USDA/land-grant system, other government offices, and agribusiness firms have the power to make many decisions that will affect the future of U.S. agriculture without regard to public opinion or, if they choose, public interest. If they use this power unwisely, dangerous biotechnological products and methods may be used by agriculturalists.

On the other hand, activists and biotechnology's opponents have powers that can cause the development of beneficial biotechnological products and methods to be unwisely neglected. Both such failures can result from an inadequate understanding of ethical issues and, perhaps, carelessness and dishonesty.

The widespread poor understanding of the wide range of ethical issues by public choice and decision makers is, itself, an important ethical issue for agriculture and rural America. Such a failure is a form of hubris and a neglect of public responsibility by persons entrusted to make decisions that will affect the future of every person born into future generations, millions of farm animals, and, perhaps, nature itself.

It can also be a double failure because inability to deal with the ethics of biotechnology can entail an inability to deal with its politics. In addition to the risk that we will do something we "ought not to do," agricultural and scientific leaders ill-informed about ethical issues will also be unable to make an effective case for beneficial biotechnological breakthroughs in science and technology in the public forum, so that we fail to do what we "ought to do" with biotechnology.

Currently, the agricultural research establishment places heavy emphasis on research to improve production techniques. Yet we neglect farm and rural problems with deficient rural institutions and policies, inadequate development of human capacity in rural areas, and our natural and manmade resources.

The agricultural research establishment has an ethical obligation to reexamine the decisions that have created and maintained these imbalances and to find a more appropriate balance. The financing of such changes was addressed in the middle section of this chapter. Still further, production research for agriculture is now swinging sharply toward greater use of the tools of biotechnology in research, a development that also requires continuous ethical assessment.

The responsibility of agricultural leaders is, first, to obtain more knowledge about the broad range of value and value-free knowledge about biotechnology's impacts, second, to be effective decision and choice makers, and, third, to be effective spokespersons in the public discussions and debates about the future of biotechnology and of agriculture itself. What is now known about the ethics of biotechnology is inadequate but still important and of potential value. The ethics of biotechnology, like biotechnology itself, is a topic for the future.

References and Additional Readings

Achinstein, P. and S. F. Barker, eds. 1969. The legacy of logical positivism. The Johns Hopkins Press, Baltimore, MD.

Arrow, Kenneth J. 1963. Social choice and individual values. 2nd ed. John Wiley and Sons, New York, NY.

Berry, Wendell. 1977. The unsettling of America: Culture and agriculture. Sierra Club Books, San Francisco, CA.

Bolognesi, Dani P. 1989. Progress in vaccines against AIDS. Science 246:1233-1234.

Briemeyer, Harold. 1965. Individual freedom and the individual organization of agriculture. University of Illinois Press, Urbana, IL.

Brill, Winston. 1985. Safety concerns and genetic engineering in agriculture. Science 227:381-384.

Brill, Winston. 1988. Why engineered organisms are safe. Issues in Science and Technology 4(3): 44-50.

Bromley, Daniel W. 1989. Economic interests & institutions. Basil Blackwell, Inc., New York, NY.

Buchanan, James and Gordon Tullock. 1962. The calculus of consent: Logical foundations of constitutional democracy. University of Michigan Press, Ann Arbor, MI.

Chargoff, Ervin. 1976. On the dangers of genetic meddling. Science 192:938-940.

Comstock, Gary, ed. 1987. Is there a moral obligation to save the family farm? Iowa State University Press, Ames, IA.

Davis, Bernard W. 1987. Bacterial domestication: Underlying assumptions. Science 235:1329-1335.

Emerson, Ralph Waldo. [1870] 1904. Farming, society and solitude. p. 139-154. Reprint. Houghton Mifflin, Boston, MA.

Friedman M. and L. J. Savage. 1948. The utility analysis of choices involving risk. Journal of Political Economy 56:279-304.

Harbaugh, Dennis, et al. 1987. Is there a moral obligation to save the family farm? Gary Comstock (ed.) p. 32-37. Iowa State University Press, Ames, IA.

Hightower, James. 1978. Eat your heart out. Crown Books, New York, NY.

Johnson, Glenn L. 1987a. Holistic modeling of multidisciplinary subject matter and problematic domains. Systems Economics, D. G. Miles (ed.) Iowa State University Press, Ames, IA.

Johnson, Glenn L. 1987b. Risk aversion vs aversion for losses and risk preference vs preference for gain. Series G-Agricultural Economics. Polish Agricultural Annual, Roczniki Nauk Rolniczych, Tom 84-Zeszyt 2, Polish Academy of Sciences. Warsaw, Poland, p. 109-119.

Johnson, Glenn L. 1988. Technological innovations with implications for agricultural economics, agriculture and rural areas approaching the twenty-first century. p. 82-108. In *R. J. Hildreth, et al. (eds.) Challenges for agricultural economics. Iowa State University Press, Ames, IA.*

Johnson, Glenn L. and Sylvan Wittwer. 1984. Agricultural technology until 2030: Prospects, priorities, and policies. Special Report 12. Michigan Agricultural Experiment Station, Michigan State University, East Lansing, MI.

Kaplan, Abraham. 1968. Positivism. p. 389-395. In *International Encyclopedia of the Social Sciences, Vol. 12.*

Lewis, C. I. 1955. The ground and nature of the right. Columbia University Press, New York, NY.

MacLean, Douglas, ed. 1988. Risk and consent: Philosophical issues for centralized decisions. In *Values at risk. Rowman and Allenheld, Totawa, NJ.*

Madden and Thompson. 1987. Ethical perspectives on changing agricultural technology in the U.S. Notre Dame Journal of Law, Ethics and Public Policy 3:85-116.

North, Douglass. 1987. Institutions, transaction costs and economic growth. Economic Inquiry, 25.

Ostendorf, David L. 1987. A protestant, populist view. In *Gary Comstock (ed.) Is there a moral obligation to save the family farm? Iowa State University Press, Ames, IA.*

Nozick, R. 1974. Anarchy, state and utopia. Basic Books, New York, NY.

Rawls, John. 1971. Theory of justice. Belknap Press of Harvard University Press, Cambridge, MA.

Rifkin, Jeremy. 1983. Algeny. Viking Press, New York, NY.

Rifkin, Jeremy. 1985. Declaration of a heretic. Routledge and Kegan Paul, Boston, MA.

Tversky, Amos and D. Kahneman. 1981. The framing of decisions and the psychology of choice. Science 211:453-458.

Williamson, Oliver E. 1985. The economic institutions of capitalism. Free Press, New York, NY.

CHAPTER TEN

Biotechnology is fraught with concern and controversy within both the scientific community and the broader public. Significant portions of the public have acquired some interest in and understanding of the new technologies and have articulated specific concerns about their applications.

The public's interest in and knowledge of biotechnology, in general, and agricultural biotechnology, in particular, have increased throughout the 1980s. While there is concern about genetic manipulation, widespread acceptance of a variety of applications exists when no risk to humans or the environment is involved. For example, a majority approve of the environmental use of genetically engineered organisms and field tests in their own communities. Moreover, roughly two-thirds of the public believe that genetic engineering will be more beneficial than detrimental.

Public enthusiasm, however, has declined since the early 1980s. Two-thirds of the American public see a likelihood of environmental risk from genetically altered bacteria, and nearly half see a similar risk from genetically engineered plants and animals. Even greater concerns are expressed about food safety, with more than three-quarters of the public expressing concern about eating genetically engineered food. Similarly large percentages of the population appear worried about the negative socioeconomic consequences for farmers. Finally, while most believe application of these new technologies to plants and animals is appropriate, a quarter believe genetic manipulation is morally wrong. Consequently, the majority of the public believes that, while research should continue, strict regulation is necessary.

The diverse set of concerns over a wide range of issues revealed in this chapter underscores the importance of developing a meaningful dialogue and mutual respect among all parties interested in or affected by these new technologies. The goal of this dialogue should not be public acceptance but public understanding, not the application of biotechnology but the development of responsible decisions regarding biotechnology applications which are compatible with community goals and values.

10 Public Perceptions of Agricultural Biotechnology

William B. Lacy, Pennsylvania State University
Lawrence Busch, Michigan State University
Laura R. Lacy, Pennsylvania State University

Introduction

The application of certain new biotechnologies to agriculture is likely to precipitate a series of technical and related social and philosophical changes of far-reaching significance for science, agriculture, and socioeconomic development. Dr. John Burress, Executive Director of the National Academy of Sciences' Commission on Life Sciences, notes, "I know it's an overused term, but what we are seeing is indeed a revolution" (Lacy, et al., 1988, p.4).

Others in government, university, and industry espouse similar predictions and perceptions. The Congressional Office of Technology Assessment (1987) views the U.S. at the brink of a new scientific revolution that could change the lives and future of its citizens as dramatically as the Industrial Revolution two centuries ago or today's computer revolution.

As the president of a large U.S. public university noted, "our society has moved into the era of high technology. In this last decade of the century, we will see more technological changes than we have experienced over the entire history of our nation. It promises to be one of the most exciting and challenging times in the history of mankind." He concluded that "biotechnology, genetic manipulation, and genetic engineering will have tremendous impact on the crops and animals we grow for food, affecting agriculture in ways never before dreamed possible" (Lacy, et al., 1988, p.4).

Despite this optimism, most of the products, processes, and impacts of biotechnology, particularly in agriculture, remain promises for the future. Moreover, the applications of biotechnology are fraught with concern and controversy within both the scientific community and the broader public. In the mid-1970s scientists intensely debated issues surrounding laboratory experimentation with recombinant DNA, expressing strong disagreements about the dangers. From this debate, scientists developed voluntary guidelines for research which were later adopted by the National Institutes of Health (NIH).

Public concerns often have centered on health and environmental safety issues. As a result, delays in field trials and product introduction have occurred. For example, an experiment to evaluate the effectiveness of a genetically engineered bacterium in reducing the susceptibility of crops to frost damage was delayed for four years. More recently (1990), Governor Thompson of Wisconsin approved legislation that temporarily banned the sale or use of a genetically engineered hormone, bovine somatotropin, in dairy cows. This action was the first prohibition of a product of genetic engineering anywhere in the country. The ban was a political compromise between forces favoring a permanent ban and those opposing any ban.

In Europe, public concerns about both the safety and the ethical aspects of genetic engineering have led to similar bans. The European debate over biotechnology has been influenced by a growing public dismay over accidents and failures of other technologies, including the nuclear power plant failure at Chernobyl, the Union Carbide accident at Bhopal, and major chemical spills in the Rhine.

For example, in Germany in 1988, an amendment to existing environmental legislation required that proposals for all new production facilities using genetically engineered organisms, whether they were inherently pathogenic or not, be subject to public debate. Moreover, the powerful Green Party is demanding that a moratorium be placed on all industrial applications of genetic engineering until more is known about the potential risks.

The European Common Market has instituted a ban on growth hormones in food products. Recently, the Advocate General of the Court of Justice of the European Communities released his opinion on the legality of the hormone ban, stating that "it was appropriate and justifiable to prohibit the administration of the five substances for fattening purposes, even in the absence of scientific evidence showing that they were harmful. A total prohibition was the only solution which could bring an end to the distortions of competition and barriers to intra-Community trade in meat, eliminate risk to public health, even if they were purely hypothetical ones, and avoid a further reduction in consumption" (Weber, 1990, p. 1).

Concern is voiced by scientists, regulators, industry, and the public, in general, that all technologies, including genetically engineered organisms and products, could have adverse impacts. As a result of these concerns and experiences with previous technologies, an increasingly accepted position among technology assessment professionals is that: (1) all technologies have multiple effects; (2) many of these effects are potentially harmful and thus require conscious decisions; and (3) these critical decisions entail moral as well as scientific analysis.

Scientists dealing with the public should avoid:

- *arrogantly dismissing lay hypotheses;*
- *ignoring nonscientific points about biotechnology; and*
- *underestimating the intelligence or power of the public.*

In democratic societies, the public has an obligation and a right to shape the developments of technology. In the case of biotechnology, as we have seen, they are increasingly exercising that obligation and right. Therefore, it is important for scientists, as well as policy makers, to understand not only the opinions of the public but also the bases for these opinions.

In a sense, societal concerns and biotechnology research may be on a collision course, confronting scientists, administrators, and policy makers with difficult decisions, new challenges, and multiple opportunities. As we approach major commercial applications of biotechnology research, society will encounter technical issues of tremendous consequence and with major social and ethical ramifications. Because of previous negative experiences with commercial applications of nuclear physics and chemistry, the public will no longer unconditionally trust scientists.

Congressman George Brown (1987) urges the scientific community and industry to avoid several things in dealing with the public. First, they must avoid an air of scientific arrogance which casually dismisses those who see biotechnology as a risky business. Brown notes that in our state of incomplete knowledge about this or any other technology, any scientist who finds himself or herself unable to consider a lay hypothesis as seriously as one made by a learned peer should, for the good of this area of science, avoid public pronouncements. Debate is a two-way process.

Second, they should avoid dismissing nonscientific points, especially the ethical ones, being made about biotechnology. The perspectives of the natural sciences, economics, and the social sciences must now be tempered by ethical considerations.

Third, they should not underestimate the intelligence or the power of the lay public. Brown concludes that a shared goal should be to develop a comprehensive, cooperative, and flexible partnership between the scientific community and the general public in the area of biotechnology.

With these observations in mind, we examine the public's awareness, understanding, and perceptions of biotechnology, with particular emphasis on agricultural biotechnology. We seek to identify the content and the changes in the public's perceptions of biotechnology and the broader context in which these perceptions are embedded.

The basic research needed to adequately understand public perceptions by demographic group, region, and industrial sector has not been done, and much important research remains to be conducted. However, some effort has been made to characterize the nature of the public's perceptions of biotechnology.

This chapter begins with a broad discussion of nine dimensions of public concerns regarding biotechnology and a brief review of the ways people process information. Next, the public's perceptions and general attitudes about science and technology are examined.

The third section deals with the public's perception of biotechnology. This is followed by a more detailed discussion of its perception of agricultural biotechnology. Finally, the public's perceptions of the credibility of information sources on biotechnology and of the appropriateness of specific agencies to regulate its applications are addressed. The chapter concludes with strategies and suggestions for developing an effective and meaningful dialogue with a diverse public.

Dimensions of Public Concerns

After the first successful directed insertion of foreign DNA into a host microorganism just 17 years ago, researchers recognized the knowledge opportunities in basic molecular biology and the commercial potential of this technology in such areas as pharmaceuticals, animal and plant agriculture, specialty chemicals, food additives, energy production, and bioelectronics. A debate over the possible benefits and risks, as well as the value issues, quickly ensued and has continued into the present. As the work has moved from research to commercial application and from the laboratory to the environment, the public debate has become more vigorous.

Despite the current relatively early stages of biotechnology development and commercial application, significant portions of the public have acquired some awareness, interest, and understanding of this new technology and have articulated specific concerns about its application. Unfortunately, certain members of the scientific and science-policy community have tended to oversimplify these concerns or to dismiss them as being illogical, irrational, emotional, or the product of scientific illiteracy.

Some have argued that the most significant risk of biotechnology is to the environment.

Members of the scientific community may also contrast public perceptions (subjective) with scientific knowledge (objective). However, strictly speaking, this contrast is erroneous. Scientists, like laypersons, must rely on their perceptions in order to make sense of scientific data. Admittedly, scientific data are more carefully contrived than are those of the public at large. Nevertheless, it is more correct to contrast public perceptions of biotechnology and scientific perceptions of biotechnology.

Like the active debates among scientists beginning in the 1970s, public concerns are quite diverse and range from the simple to the complex and sophisticated. Public perceptions about biotechnology can be characterized in terms of several major dimensions or areas of concern, which include both risk and value issues. A discussion of nine of these dimensions follows.

Environment

The perceived risks of biotechnology to the environment have become a major issue. Public concern about risk associated with genetically engineered organisms is due in part to their uniqueness. These organisms are viewed as inherently more unpredictable than chemical products, since living organisms reproduce, grow, migrate, and mutate. Furthermore, while any given introduction has only a small chance of becoming a problem, each is characterized as a low-probability, high-incident risk, with long-run consequences that can be enormous and irreversible.

Some have argued that the most significant risk of biotechnology is to the environment. A portion of the public believes that the release of genetically engineered organisms may result in some unintended and perhaps permanent damage or loss to the environment. The analogy is often made to previous problematic introductions of exotic organisms into North America (e.g., kudzu vines, African bees, and gypsy moths).

They warn that engineered organisms might also become pests, displace existing plants and animals, disrupt the functioning of ecosystems, reduce biological diversity, alter the composition of species, threaten the extinction of various species, and even change climate patterns (Mellon, 1987). Therefore, we need more knowledge of the factors that affect the ability of species to survive, proliferate, and disperse in nature and a greater understanding of the potential for genetic exchange in nature between genetically engineered organisms and other species.

Health

Members of the public also have concern about the risks to human health from engineered organisms. Since companies will not normally develop for release organisms known to harm humans, the primary risks will be indirect. However, in the past, the public has been exposed to negative impacts from substances added to food and pharmaceuticals, either intentionally (e.g., food additives and food coloring) or unintentionally (e.g., pesticides and plant nutrients).

Portions of the public believe that certain genetically engineered products could have similar unintended consequences, for example by increasing levels of toxins in foods or increasing the number of disease-causing microorganisms that are resistant to antibiotics. Moreover, there are concerns that genetically engineered food products will increase further the growing complexity and product differentiation within the food industry, making it more difficult to select a nutritious diet.

Agriculture

Relative to agriculture, some specific concerns overlap the perceptions of environmental and health risks. The public has raised concerns about how biotechnology may affect plants and animals and whether it will endanger the food supply. Some have questioned whether crop plants genetically engineered for resistance to all the major herbicides may themselves become intractable weeds, resistant to all the major herbicides. Others are concerned that herbicide-resistant crops may cross with wild or weedy relatives and thereby spread genes for herbicide resistance into natural populations.

With respect to farm animals there has been concern about possible pain and suffering as a consequence of genetically engineered changes. Will dysfunctional changes in morphology, abnormal metabolism, dysfunctional behaviors, autoimmune diseases, and the early onset of degenerative changes occur as an unforeseen consequence?

Finally, there is concern that biotechnology may increase the rate of plant and animal extinction and reduce regional and global biodiversity. Consequently, a number of the general public beliefs about biotechnology are specifically applicable to agricultural applications and products. However, public concerns are not limited to risks to agriculture, the environment, and human health. In fact, a number of other important issues, such as those discussed below, may even transcend those relative to risk.

Science

A fourth issue raised by biotechnology involves the basic methods and approaches of science. The founding fathers of modern science helped to define a particular approach and method for it. In brief, their method attempted to reduce nature to small, definable pieces, subject to human manipulation and separated from broader questions of value. From this perspective, scientists could control, measure, reduce, and divide nature in order to generate that sort of knowledge which is power.

Biotechnology is viewed as just another step in this process of reductionism in science. The concern is that this approach, while providing only partial knowledge, has become the dominant epistemology, often to the exclusion of alternative ways of knowing.

Moreover, biotechnology is seen as permitting scientists to rearrange the genetic code in order to produce short-term commercial benefits for the institutions involved, while de-emphasizing a more sophisticated understanding of complex ecosystems. Many persons, both inside and outside the scientific community, have argued that current scientific practices tend to permit only the most simplistic understanding of the subcellular world.

Social Justice

A fifth dimension of public concern regarding biotechnology is the social and economic impacts. This issue is increasingly cited as a rationale for moving cautiously on new technologies. The concerns focus on questions of distributive justice, equity, and redistribution. A fundamental issue here is who will gain and who will lose in the application of biotechnology.

In agricultural biotechnology, economists have predicted, for example, that within three years of market introduction of the genetically engineered hormone, bovine somatotropin, upwards of 25-30% of all American dairy farmers will be out of business. They contend that this consequence may be the price we pay for the equilibrium adjustment in the market (Kalter and Tauer, 1987). Others have suggested that large farm operations may have a substantial advantage over small operations in benefitting from biotechnology.

Furthermore, many are concerned that biotechnology will fail to open up new opportunities in the rural economy. Will, they ask, the new technologies lead to more jobs? Where will those jobs be located? Will those jobs be more meaningful, fostering self-respect and autonomy? While biotechnology is only one factor among many contributing to social and economic developments, it has tended to stimulate debates regarding questions of equity and social justice (Busch, et. al., 1990).

Economic Concentration

On a broader level, there is concern that biotechnology will continue and accelerate the trend towards increasing concentration of power in the hands of a small number of corporations. The scale of investment in biotechnology is accelerating rapidly in Western Europe, the U.S., and Japan. (For example, commercial development of biotechnology is predicted to account for over 10% of the Japanese gross national product by the year 2000). This investment is increasingly dominated by a relatively small number of large, multinational corporations in the pharmaceutical, oil, and chemical industries.

Biotechnology has permitted corporations to diversify into every field or specialty that uses living organisms as a means of production. It has also further reduced distinctions among the traditional industrial sectors, rendering corporate boundaries virtually unlimited. This has resulted in widespread horizontal and vertical integration across national boundaries.

For example, large nonpharmaceutical corporations, such as Du Pont, Monsanto, Procter and Gamble, and Dow Chemical, have developed a new emphasis on and invested heavily in pharmaceutical efforts through biotechnology research and development. Similarly, large chemical, oil, food, and pharmaceutical corporations have taken the leadership in agricultural biotechnology and in various sectors of agribusiness (e.g., seeds, farm chemicals, food processing). It is predicted that by the next century this small number of highly diversified, multinational corporations will likely control most of the food system.

Consequently, development and commercial control will be in the hands of corporations that transcend geographic boundaries and hold limited national allegiance. Within this context, people question how we can ensure that democratic participation will occur in the decision-making processes surrounding the development and commercialization of biotechnology. This is difficult within national boundaries and generally prohibitive internationally, given current governmental structures.

Progress

The seventh concern involves the concept of progress and the notion of a technological imperative. Some have characterized those who are opposed to biotechnology as being simply antitechnology and antiprogress (i.e., latter-day Luddites). They argue that what is technologically possible will eventually be done, regardless of social, economic, and ethical issues.

Others have argued, however, that we need to redefine progress, extending it beyond terms such as output, utility, and efficiency. According to this position, a definition of progress should encompass the entire context in which human beings live, including all of the economic, social, and ethical dimensions. Biotechnology, then, needs to be evaluated in terms of its contribution to progress as it is redefined.

Nature

Another issue concerns nature itself. The question here is the appropriateness of manipulation, control, and domestication of nature to serve our economic purposes versus protection of nature from human control and manipulation. One of the concerns and goals of a portion of the public is to protect the ecological and environmental status quo from change, even those changes that would arguably benefit mankind.

Proponents of this perspective assert that nature has an order, an integrity, and a purpose of its own that one should respect, not for our sakes but for its sake. They fear that market forces, coupled with the new biotechnologies, will impose the will of economic man on all of nature, quickly destroying the little that remains of our evolutionary and ecological heritage. They also fear that the technologies which provide us with the potential to redesign nature may serve to undermine our ability to decide what kind of nature we want.

From this perspective, intentional releases and the successes of genetically engineered organisms in the environment are of greater concern than any unanticipated or unintentional consequences (Sagoff, 1988). Importantly, those concerned with preserving the integrity of nature are not posing the issue strictly in terms of risk and benefits.

Sanctity of Life

A final important public concern regarding biotechnology involves the ability to modify and manipulate living organisms. At the heart of this argument are fundamental moral and ethical issues.

Some opponents of genetic engineering have argued that humans should not "play God" by manipulating the genes of humans and other organisms, thereby creating a "second nature." Advocates of biotechnology respond that humans have manipulated genes for thousands of years through selective breeding. Opponents disagree, arguing that genetic changes through breeding have been limited and have not involved crossing fundamental species barriers. They contend that genetic engineering can render such barriers irrelevant.

People are more accepting of technologies that are:

- *voluntary;*
- *controlled by individuals;*
- *familiar;*
- *risky in known ways; and*
- *fair in terms of socioeconomic impact.*

Additionally, biotechnology makes any animal, including Homo sapiens, no different than a manufacturing process that can be claimed as an invention and patented. They argue that the sacredness of life, as well as the very concept of life, are challenged by genetic engineering.

Information Processing

In addition to the nine risk and value dimensions discussed above, the ways that people process information may also significantly affect public perceptions. For example, people tend to categorize information, including the acceptability of new technologies, in terms of a number of dichotomies. Furthermore, people cognitively process information differently based, in part, on personal experiences and learning styles.

Scientists have long posited that humans wish to believe they have freedom over their actions and power over their environment. When developments, such as new technologies, appear to threaten their freedom and control, people tend to react negatively. Therefore, people are more likely to accept new technologies if they view their introduction as voluntary rather than involuntary. They are more accepting of them when they see the technologies as controlled by themselves (e.g., seat belts, helmets) versus those controlled by the system (e.g., nuclear power plants). People are also more comfortable with those technologies that they see as familiar as opposed to those that are exotic and futuristic (e.g., biotechnology).

Another dichotomy that affects the acceptance of new technologies is whether the risks and benefits are known or unknown. Even when very low levels of unknown risks are reported, the public may be less willing to accept them than higher levels of known risks. Finally, the public is more receptive to technologies they believe are fair versus those that are unfair in terms of their social and economic impact. People tend to assess whether the risks and the benefits are going to the same people. They also ask if the risks are distributed according to an acceptable decision-making process (Sandman, 1987).

One of the more consistent and pervasive results of research on public perceptions of technology is that different individuals interpret the consequences of technologies in different ways. For example, benefits and risks have different meanings in different contexts. In addition, there is little correlation between the definitions of risk used by experts and the definitions applied by ordinary people to the benefits and hazards they experience.

Therefore, the expectation that better communication about risks and benefits will alleviate the conflict may be erroneous. First, the same communication may be interpreted in different ways by different individuals and social groups, and second, the same information may be evaluated differently by different people (Cannell and Otway, 1988). Finally, these disparate ways of information processing exist within the complexity of the public's broad and diverse world views, value systems, and variable life situations.

Within this context, we next review surveys of public knowledge and opinion about issues concerning science and technology, in general, and genetic engineering and biotechnology, in particular. These studies, particularly those conducted by the National Opinion Research Center, the National Science Foundation, and the Office of Technology Assessment, utilized sophisticated sampling procedures which produce an unbiased estimation of perceptions and opinions of the adult population of the U.S. While survey results are subject to sampling error, the range of error for these samples is generally less than 3%.

Perceptions of Science and Technology

To better understand the public's perception of biotechnology, it is important to position these perceptions in the broader context of the public's perception of science and technology. Studies to assess public opinion on science, technology, and public policy have been conducted by the National Opinion Research Center since the mid-1950s and by the National Science Foundation since the early 1970s. These have been complemented by studies in the 1980s by the Office of Technology Assessment, the Public Opinion Laboratory of Northern Illinois University, North Carolina State University, the Nova Laboratories, Inc., and other organizations. Only a brief summary of results from these studies is presented here. For more detail the reader is referred to the original reports.

Interest and Understanding

Despite differences in conceptual framework, as well as questionnaire wording, various studies in this area reach the same general conclusion. For example, in the Office of Technology Assessment (OTA) study approximately a quarter of the adult population of the United States reported being very interested in science, and a smaller minority (16%) reported having a very good understanding of science and technology. However, a clear majority reported being somewhat interested and adequately informed about science, technology, and related public policy (OTA, 1987).

The respondent's education level clearly affects the self-reported interest in and understanding of science. For example, 40% of the college graduates reported being very interested in science, and 29% of this group rated their science understanding as very good (OTA, 1987). However, the public's interest in and understanding of science appear to have declined between 1982 and 1986, a change which reflects a large decline in understanding and interest among those without a college education.

A portion of the U.S. population also actively seeks scientific information. One quarter of the public reported reading books or magazines on science and technology, either daily or weekly. In addition, 17% of the public reported that they have, or someone else in the household has, a science- or technology-related job. This public interest and exposure to science issues is not limited to people who report understanding science. Forty percent of the adults who feel they have only adequate understanding of science and 20% who feel their science understanding is poor reported reading the science section of the newspaper at least weekly.

This interest in science issues in the United States is not restricted to the experts, although it is far more common among the better educated. Finally, nearly one-third of the American public reported that they are very concerned about science policy, and another 50% said that they are somewhat concerned (OTA, 1987).

Beliefs About Science

Most studies show that a majority of the public, regardless of their understanding and interest in science, believe that the benefits of scientific research outweigh the harmful results and that science generally makes our lives better. Although the public indicates that it expects some risks from scientific and technological development, a large majority still believe that the benefits to society from these innovations will outweigh the risks. Furthermore, the risks of scientific and technological development are frequently viewed as overstated (OTA, 1987; Berrier, 1987; Russell, et al., 1987).

Despite the basically positive orientation of the public toward scientific and technological developments, there have been significant shifts in attitudes over time. These shifts have resulted in public support for increased control of technological development.

The National Opinion Research Center surveys between 1957 and 1979 indicated a significant decline in the number of respondents reporting that the benefits of science outweigh the harmful results, particularly among those respondents who reported a low interest or understanding of science. In 1957, 87% of this group reported perceiving the benefits to outweigh the harmful results, while in 1979, only 66% so indicated. While those reporting greater interest in and understanding of science had considerably more positive attitudes, their pattern of change between 1957 and 1979 was very similar (Berrier, 1987).

The OTA report confirmed this perception of science in the U.S. The OTA survey found that 41% of Americans said that they expect a lot of benefit for themselves and their families from the development of science and technology over the next two decades, with nearly an equal number (39%) indicating they expect some benefit. Public expectations concerning the benefits of science increase with education, from 28% of those with a high school degree to 57% of college graduates saying they expect a lot of benefit (OTA, 1987). However, this may reflect not only an increasing appreciation of the benefits through education, but an increased likelihood of benefits because of position in society.

These perceived benefits do not come without some perception of potential risks. Slightly more than one-fifth (22%) felt that advances in science and technology cause a lot of risk to themselves and their families, while nearly one-half (49%) believed these developments will pose some risks.

In sharp contrast to its influence on the expectation of the benefits, education does not appear to influence the expectation that a lot of risk will be caused by scientific and technological innovation. Concern about the personal risks of scientific and technological development appears to be uniform across most subgroups of the U.S. population. Neither age, education, nor interest in and understanding of science seem to significantly affect concern about the risks of science and technology (OTA, 1987). Consequently, education may increase appreciation of the benefits of science, but it is not likely to decrease perception of the risks.

When asked to assess the relative benefits and costs, a majority of the American public (62%) reported believing that the benefits of continued scientific and technological development will outweigh the related risks. Importantly, the public appears to have mixed opinions about the risks of scientific and technological development.

A sizeable majority of the total population (61%) said they agree with the proposition that "society has only perceived the tip of the iceberg with regard to the risks associated with modern technology." A majority (59%) also agreed with the proposition that "the risks associated with advanced technology have been exaggerated." Additionally, a significant minority (25%) believed that the rate of growth in science is either much too fast or a little too fast, and 42% of the public agreed with the notion that "unless technological development is restrained, the overall safety of our society will be jeopardized significantly in the next 20 years" (OTA, 1987).

A majority of the public still appear comfortable with the present degree of regulation and control of science and technology. However, a large minority (43%) believed that the degree of control society has over science and technology should be increased. Analysis of opinion changes over time suggests that the demand for stricter controls may increase. Comparing National Science Foundation survey results from 1972, 1976, and 1986-87 reveals that the percent favoring expanded control increased from 28% to 31% to 43%. Similar shifts over the next decade would result in a majority of the public favoring increased control (OTA, 1987).

In summary, Americans remain generally optimistic about the benefits of scientific and technological development, and they believe that the benefits generally outweigh the risks. Confidence in science and technology, however, has declined over the last three decades. This, coupled with a substantial and growing concern about the risks and unrestrained growth of technology, has led Americans to favor expanded regulation of scientific and technological development. It is within this context that we now examine the public's perceptions of biotechnology and genetic engineering.

Perceptions of Biotechnology and Genetic Engineering

Citizens in the U.S. are being exposed increasingly to information about biotechnology and, particularly, about genetic engineering. A limited number of surveys have been conducted within the last five years to assess the public's awareness, understanding, and perception of the impact of genetic engineering, the most comprehensive being the national survey by the Office of Technology Assessment (1987).

Interest in and Understanding of Biotechnology

Public perceptions of biotechnology will be shaped in part by public interest in and knowledge of the issues. Public interest is low to moderate at this point but increasing. A combination of interest in science and media exposure has resulted in an American public that is becoming aware, although not necessarily sophisticated, about biotechnology.

Surveys conducted from 1981 through 1989 found levels of interest and knowledge about biotechnology ranging from slightly less than 50% in the early 1980s to a high of 85% near the end of the decade. For example, in 1985, face-to-face interviews with 2,000 adults nationwide revealed that 49% of the public reported at least some awareness of biotechnology (Berrier, 1987). Two years later, in another national study, three in five (61%) members of the public reported having heard of genetic engineering (Russell, et al., 1987).

Similarly, late in 1986 the OTA national survey found moderate awareness of genetic engineering among the American public. Less than a quarter of the public (24%) reported that they had heard or read almost nothing about genetic engineering, while more than a third of Americans (35%) said that they had heard a fair amount or a lot about genetic engineering.

Education is a key factor in awareness, with 26% of high school graduates and 61% of college graduates reporting that they had heard or read at least a fair amount about the topic. There also may be regional differences in awareness. In a 1989 study in North Carolina, slightly less than two-fifths reported that they had heard or read either a lot (6%) or some (32%) about genetic engineering (Hoban, 1990). In contrast, 58% of a selected rural Pennsylvania sample reported having heard or read about genetic engineering (W. Lacy, 1990, unpublished data).

Surveys also have attempted to assess respondent understanding of and general interest in genetic engineering and biotechnology. A national survey in late 1985 found that only 16% of the general public reported a clear understanding of recombinant DNA, with another 27% reporting a general sense of genetic engineering technologies (Berrier, 1987). A year later an OTA report went further and asked people to describe what was meant by genetic engineering. The results suggest that more than half of American adults (56%) can provide a meaningful, though not necessarily strictly accurate, explanation of genetic engineering (OTA, 1987).

Respondents were also asked to indicate whether they understood the meaning of a variety of terms common to biotechnology. It is important to note that in a general survey, respondents tend to overestimate their understanding of vocabulary. In this survey, 85% of the public said they understood the meaning of the term "gene," while nearly three-quarters (73%) said they understood the meaning of "chromosomes." Interestingly, "DNA," a term virtually unknown outside research laboratorie[illegible] was reported to be understood by 92% of the population.

Importantly, the best predictor of self-reported understanding of genetic engineering and biotechnology is the degree of exposure to the information, with nearly all (93%) of those who indicated they had heard at least a fair amount about genetic engineering also indicating they understood it (OTA, 1987). While these findings do not demonstrate that two-thirds of the American public really understand the meaning of genetic engineering, the survey does indicate that a substantial number of Americans believe they understand the terminology and concepts of genetic engineering and biotechnology.

Social Acceptability of Biotechnology

Public acceptance of the genetic manipulation of human cells, other animal cells, plant cells, and bacteria was assessed in the OTA survey (1987). The results indicate that the public clearly differentiates among types of organisms in stating their degree of acceptability for genetic manipulation.

On a 10-point scale, the acceptability of the genetic manipulation of human cells in the laboratory fell slightly below the midpoint (4.5), on the unacceptable end of the scale. In contrast, the public believed that genetic manipulation of other animal cells and of bacteria was more acceptable, 5.3 and 5.6, respectively. Finally, genetic manipulation of plant cells received the highest level of public acceptance (6.6). The effect of religiousness on the acceptance of genetic manipulation was marked and significant. Those who were very religious ranked each of the manipulations as more unacceptable than did those who were not too religious.

Another important question examined whether public acceptance of biotechnological applications was rooted in the end objectives or the living organism manipulated. Respondents were asked, first, to assume that the applications involved neither direct risk to humans nor an environmental risk and then to evaluate their approval of different applications of genetic engineering. Seven applications of genetic engineering were examined: a new treatment for cancer; new vaccines; cures for human hereditary diseases; disease-resistant crops; frost-resistant crops; more productive farm animals; and larger game fish.

A clear majority of the U.S. public approved of all seven applications of genetic engineering in the survey, ranging from a high of 96% approval for new treatments of cancer to a low of 66% to produce a larger game fish (Figure 1). The results also illustrate a variation in the enthusiasm for financing these applications of genetic engineering, with 75% strongly approving of new vaccines and only 25% strongly approving of larger game fish (OTA, 1987). The uses with the most immediate human benefits were at the top of the list.

A 1985 Harris poll (Berrier, 1987) and the 1987 Nova Laboratories Report (Russell, et al., 1987) also reveal strong support for genetic treatment to cure or prevent disease. However, there is very little support for genetic treatment to

Figure 1. *Levels of Approval of Different Applications of Genetic Engineering.*

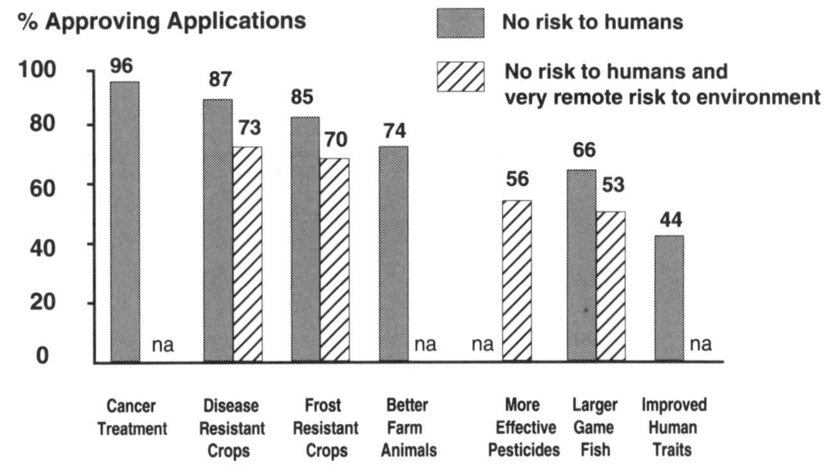

na = Comparative data unavailable
Source: Office of Technology Assessment Survey (OTA, 1987)

improve the physical or mental traits of one's children. Eighty-eight percent of the population felt that it was inappropriate for parents to have their genes changed to assure that their children would be smarter, physically stronger, or better looking (Berrier, 1987).

In the OTA survey (1987), a majority of Americans said that it was not acceptable to change the makeup of human cells. A substantial minority (42%), however, felt that it was morally wrong. Paradoxically, a majority of Americans who think human gene manipulation is morally wrong in the abstract approve it when it is used to save lives and heal sick children. For example, 84% of the American public approved changing the makeup of human cells to stop children from inheriting a usually fatal genetic disease (OTA, 1987). Thus, relative to the public's perceptions about human gene therapy, almost all Americans, regardless of age, race, education, religiousness, and even moral reservations about genetic engineering, approve and are willing to use these therapies to save lives.

Finally, a significant minority (33%) agreed that it would be better if we did not know how to genetically alter cells at all (OTA, 1987). Agreement with this statement was highest among with those without a high school diploma (43%) and lowest among college graduates (19%). In contrasting these responses with the more general anti-science sentiment reflected in an opposition to meddling with nature (24%), there seems to be greater opposition to a biotechnology-specific capability to alter cells (33%).

However, it should be noted that a majority disagreed with each of these sentiments and favored science and biotechnology. A majority of the public also indicated that they thought that the risks of genetic engineering have been greatly exaggerated (55%) and that unjustified fears of genetic engineering have seriously impeded the development of valuable drugs and therapies (58%). Finally, genetic engineering and biotechnology were much more popular when the public was queried about specific instances rather than in the abstract (OTA, 1987).

Impacts of Biotechnology

As noted earlier, the American public has a positive orientation toward science, and this is reflected, as well, in their attitudes toward biotechnology and genetic engineering. However, the degree of positive reaction to science varies widely over specific topics.

In the surveys of the early 1980s, only 5% of the population thought science and technology will do more harm than good, but three times that proportion of Americans saw the risks of biotechnology outweighing the benefits (Berrier, 1987). Just a few years later, 28% of the public responded that the benefits of science and technology will not outweigh the risks (OTA, 1987).

However, when participants were asked whether they thought each of five scientific developments (solar energy, organ transplants, genetic engineering, robots and automation, and nuclear power) would make life better or worse for people, there was a wide range of opinions. At the high end of the scale, nearly everyone (92%) felt that solar energy will make the quality of life better, and 87% believed that organ transplants will do the same. In contrast, only 60% in the case of robots and automation, and about half (51%) in the case of nuclear power felt that they will make life better. Opinions on genetic engineering fell between these two levels, with two-thirds of the public saying they they believed it will make life better (Figure 2) (OTA, 1987).

These results are similar to those of a 1987 survey by Nova Laboratories, Inc., which showed that two-thirds of the respondents believed that they or their family will receive benefits from genetic engineering within their own lifetime. Interestingly, a larger majority (87%) believed that people around the world will benefit from this technology. This study concurred with the OTA results concerning perceptions of genetic engineering in terms of safety. When respondents were asked to rate the safety or danger of agriculture, nuclear energy, and genetic engineering, agriculture was rated as the safest of the three areas (52%), with genetic engineering intermediate at 26%, and nuclear energy at 15%. Agriculture was rated as more dangerous than safe by 26%, genetic engineering was so rated by 41%, and nuclear energy by 69% (Russell, et al., 1987).

In contrast, the acknowledgement of actual dangers is relatively low. In the OTA survey, only 19% of the public indicated that they had heard of any potential dangers from genetically engineered products. This awareness rose with education, with 13% of those with a high school degree compared to 34% of those with a college degree reporting having heard of the dangers. An even smaller percentage could specify a potential danger (12%), with the most cited being contamination and spread of potential products, health hazards and mutations, and environmental contamination (OTA, 1987).

Despite the low level of knowledge about actual dangers from genetically engineered products, half of the public (52%) stated that they thought that it was somewhat likely or very likely that genetically engineered products will represent a serious danger. This general sense that a serious danger from

Figure 2. *Opinions Regarding Whether Particular Scientific Developments Will Make Life Better.*

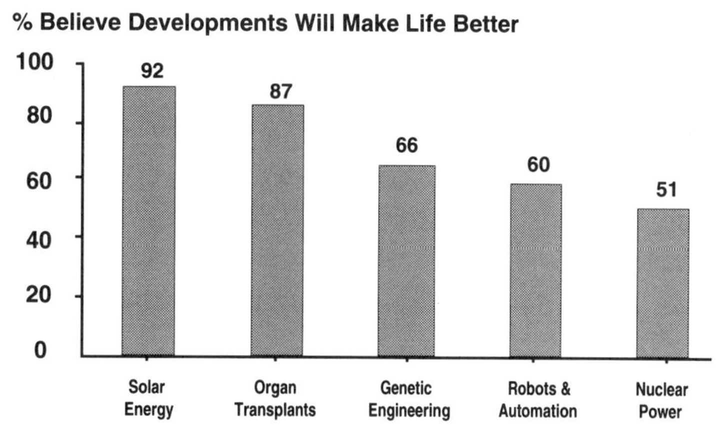

Source: Office of Technology Assessment Survey (OTA, 1987)

genetically engineered products is at least somewhat likely exists independent of education or information about the products (OTA, 1987). Similarly, the Nova Laboratories study reported that four in ten described genetic engineering as being risky. Interestingly, however, those who believed genetic engineering was safe and beneficial were more likely to be male, under 45, and better educated (Russell, et al., 1987).

Attitudes regarding biotechnology vary among leadership groups. Only 2% of the science policy leaders saw the risks of genetic engineering as the single most important public policy issue involving science and technology, while 9% of the environmental group leaders took this position. Moreover, while 17% of the science policy leaders said that the conduct of experiments involving genetic engineering was a major problem, 42% of the religious leaders took this position. According to this study, however, science policy, environmental organization, and religious leaders all indicated that they saw the benefits of biotechnology as being greater than the risks (Berrier, 1987).

There appear to have been some changes in these perceptions from the early to mid-eighties. First, the proportion of Americans believing that genetic engineering will make life a lot better has declined from 32% in 1982 to 18% in 1986. Second, the proportion who think genetic engineering will make life worse increased from 16% in 1982 to 22% in 1986 (OTA, 1987). In a 1989 survey of North Carolinians, the percentage who felt that the risks will be greater than the benefits was 29% (Hoban, 1990). Thus, while a substantial majority of Americans still believe genetic engineering will make life better rather than worse, these surveys found that the public's enthusiasm about the benefits of genetic engineering has declined since 1982.

Regulation of Biotechnology

Given the mixed feelings about biotechnology among the American public, a number of questions have arisen. While a large majority of the American public (82%) believed that genetic engineering and biotechnology research should be continued, more than three-fourths of the public (77%) agreed with the statement that the potential danger from genetically altered cells and microbes is so great that strict regulations are necessary. Indeed, 43% reported that they agree strongly with this statement (OTA, 1987).

While the public acknowledges that unreasonable fears may have delayed significant benefits from this technology, it still perceives potential dangers from the innovations and from the scientists themselves. Therefore, the public believes that strict regulation is necessary. For example, a majority of the public agreed with the statement that because of their knowledge, scientific researchers have a power that makes them dangerous (Berrier, 1987). Similarly, the Nova Laboratories survey reported 70% of the public favoring limitations on scientists regarding the kind of genetic engineering research they should be allowed to conduct (Russell, et al., 1987).

The 1989 survey of North Carolina residents (Hoban, 1990) also asked several questions regarding regulation. The respondents were nearly unanimous (97%) in agreeing that regulation of biotechnology and genetic engineering should cover public health risks and potential environmental damage. In addition, almost as many respondents (87%) thought that regulations should cover adverse economic impacts on farmers. Furthermore, three-quarters of the respondents believed that regulations should somehow try to limit potential social impacts on rural communities.

A key issue is the proper balance of regulation to provide the necessary protection and to avoid excesses. When asked about too little regulation posing serious risks to human health, 84% agreed with the statement. On the other hand, 59% agreed that too much regulation of genetic engineering will hurt American economic competitiveness.

Finally, opinions were assessed on the source and adequateness of regulation and the participants to be involved in regulation-policy decision making. Over four-fifths (81%) indicated that the states should have their own regulations, and just half (52%) felt that federal regulations are adequate. Moreover, the respondents believed that the public should have more say in government regulation of genetic engineering (Hoban, 1990).

Conclusion

The public's interest in and knowledge of biotechnology have increased throughout the 1980s, with approximately half the population now indicating moderate awareness of genetic engineering. While there is concern about genetic manipulation, particularly of human cells, widespread acceptance of a variety of applications exists when no risk is involved to humans or the environment. Moreover, roughly two-thirds of the public believe that genetic engineering will be more beneficial than detrimental. However, public enthusiasm has declined since the early 1980s. In addition, about half the public believe genetically engineered products represent a serious danger. Consequently, the public believes research should continue, but strict regulation is necessary.

Perceptions of Agricultural Biotechnology

The public's concerns about agricultural biotechnology are greatly influenced by their broader concerns about science and technology and their general attitudes regarding agriculture and genetic engineering. However, important additional factors also affect their perceptions of agricultural biotechnology, including environmental risks and food safety, the use of products which require deliberate release of genetically engineered organisms into the environment, and the socioeconomic impacts.

The U.S. population is moderately aware that genetic engineering is being applied to agriculture and is used to produce altered plants and animals. For example, four out of ten Americans (41%) reported that they have heard about gene splicing or recombinant DNA to produce hybrid plants and animals. This awareness ranged from a low of 29% among those with less than a high school degree to a high of 62% among those with a college degree (OTA, 1987).

Environmental Impacts of Agricultural Biotechnology

The application of biotechnology to the area of agriculture has raised several questions and concerns regarding the environment. Research and product development in this area require the deliberate release of genetically engineered organisms into the environment and, consequently, have stimulated increasing concerns about environmental risks. These concerns are part of a larger and broader environmental movement which began in the 1960s and early 1970s. To better understand the public's environmental concerns about agricultural biotechnology, it is important to be aware of its broad concerns about environmental quality.

Concerns

It appears that agricultural biotechnology will be carefully scrutinized regarding its potential adverse environmental impacts. The American public indicates that it desires strong environmental protection, and this preference is growing.

In a national CBS/New York Times poll in 1981, 45% agreed that protecting the environment is so important that requirements and standards couldn't be too high and that continuing environmental improvements should be made regardless of cost. Just five years later in a similar CBS/New York Times national poll, 68% (an increase of 23%) agreed with the same statement (Berrier, 1987). By 1989, a survey of North Carolinians (Hoban, 1990) revealed that almost all respondents thought companies (97%) and the government (80%) should spend more money to protect the environment. Moreover, they thought that the government should protect the environment even if it limited economic development (85%). Finally, almost two-thirds indicated a willingness to pay higher taxes for more effective environmental protection.

In the OTA survey (1987), a third (32%) of the U.S. public thought the overall quality of the environment is "getting better compared to ten years

Almost all Americans think that the government and industry should spend more on environmental protection.

Figure 3. *Familiarity with and Concerns About Environmental Issues.*

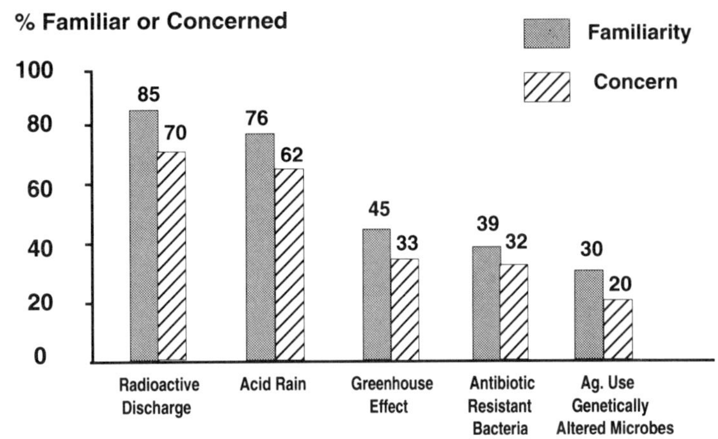

Source: Office of Technology Assessment Survey (OTA, 1987)

ago." However, the largest group of respondents (39%) believed the overall quality of the environment is getting worse. Overall, two-thirds of the American public believed that the quality of the environment has been unchanged or deteriorating during the past ten years. A couple of years later the percentage perceiving environmental degradation had apparently increased. In a North Carolina survey, over half (55%) of the respondents indicated that air and water quality in their state is worse now than ten years ago (Hoban, 1990).

To assess public awareness of associations between technology and adverse environmental consequences, the OTA survey presented five types of environmental problems with a possible technological origin: radioactive discharge from nuclear power plants; acid rain; the greenhouse effect; antibiotic-resistant bacteria; and agricultural use of genetically altered microbes. Familiarity with these issues ranged from highs of 85% for radioactive discharge and 76% for acid rain through 45% for the greenhouse effect and 39% for antibiotic-resistant bacteria to a low of 30% for the agricultural use of genetically altered microbes (Figure 3). While education and science understanding did increase public recognition of these environmental issues, significant portions of all groups in the population indicated that they had heard of these issues (OTA, 1987).

Respondents were also asked about their level of concern on these five environmental issues (OTA, 1987). Once again there was a wide range of opinion, with the greatest concern being expressed about radioactive discharge (70% indicating very or somewhat concerned) and acid rain (62% indicating very or somewhat concerned). Paralleling the levels of awareness, much lower levels of concern were expressed about the greenhouse effect (33%), antibiotic-resistant bacteria (32%), and agricultural use of genetically altered microbes (20%) (Figure 3).

These latter relatively low levels of public concern, while accurate at the time of the survey, are probably misleading in the long term. Public concern is likely to grow and has indeed grown since 1986. With increasing awareness will likely come increasing concern.

Effects

Evaluation of public perceptions regarding the impact of technological developments on the environment indicates that approximately a third (32%) believed that it has a negative impact. This negative assessment appears to be unrelated to age, education, or science orientation. Interestingly, concern about environmental effects of technology appears to be unrelated to the perceived risk-benefit tradeoff of scientific growth.

While the public's concerns regarding genetically engineered products are not well articulated, the public differentiates among organisms when asked the likelihood of genetically engineered organisms in the environment producing specific negative outcomes. In the OTA survey (1987), the public was asked about seven potential dangers of environmental release. The dangers from using genetically engineered organisms in the environment perceived most probable were: production of harmful, genetically altered bacteria (68%); the creation of antibiotic-resistant diseases (61%); the creation of herbicide-resistant weeds (56%); and the endangerment of the food supply (52%).

In addition, a large minority of the public believed that it is likely that the environmental release of these organisms will increase the rate of plant or animal extinction (45%) and change rainfall patterns (42%) (Figure 4).

Moreover, the likelihood of environmental risks from genetically altered plants and animals was seen as very likely (13%) or somewhat likely (34%) by the American public. In all these perceptions, however, the risks were generally perceived as somewhat likely rather than very likely.

Given the perception that risks from genetically altered organisms pose a likely threat to the environment, the issue of acceptable levels of risk, if indeed any level is acceptable, becomes a related question. In the OTA survey (1987), respondents were asked if they would approve use in the environment of an organism which would significantly increase farm production, with no direct risk to humans, but a risk of losing some local species of plants or animals. Respondents were asked initially whether they would consider a one in one hundred level risk and, if not, would they approve subsequent specific risk levels up to one in one million. They were also asked if they would approve if the risks were unknown or known but very remote. (While a debate exists about the public's ability to meaningfully interpret probabilities, the responses do provide an indication of perceived levels of acceptable risk.)

The survey revealed that the public is not risk adverse, at least if the risk is restricted to a local ecological disruption. While only 40% approved at a one in one hundred risk level, a majority (55%) did approve environmental use of genetically engineered organisms if the risk of local plant or fish extinction was no more than one in one thousand. This rate of approval increased to 74% at a risk of one in one million. However, even at this remote level of risk, nearly a fifth of the population (18%) said they do not approve of the environmental application of genetically engineered products (OTA, 1987).

An important finding in this research is the general rejection of unknown risks. If the public perceives the risk as truly unknown, only 31% said that they approve of the environmental application. In fact, more people approved of the application at a relatively high level of known risk, such as one in one hundred (40%). Even when the survey asked about an unknown but very remote risk, only 45% of the public approved the environmental release. This suggests that the public evaluates an unknown, but very remote risk as somewhere between one in one hundred and one in one thousand (OTA, 1987).

Finally, the perception of even very remote environmental risk seems to affect the willingness of the public to approve environmental uses of a range of genetically engineered products. Public acceptance of these products is very high when it is indicated that there is no direct risk to humans and no reference to any environmental risk is made. However, when OTA asked the public its opinion about environmental uses of specific genetically engineered products under remote risk conditions to the environment, they found the following reductions in approval rates: disease-resistant crops (87 to 73%); frost-resistant crops (85 to 70%); and larger game fish (66 to 53%) (Figure 1) (OTA, 1987).

Figure 4. *Likelihood of Specific Dangers from Use of Genetically Altered Organisms in the Environment.*

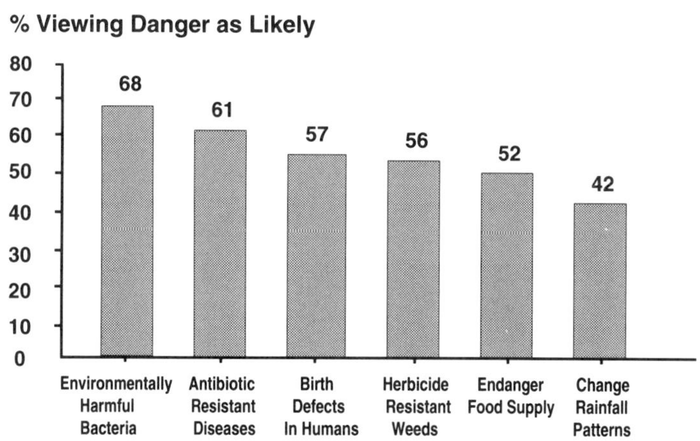

Source: Office of Technology Assessment Survey (OTA, 1987)

Field Testing

One of the most pressing issues of agricultural biotechnology has been the field testing of genetically engineered organisms. The first small-scale field trials of genetically engineered microorganisms took place in the United States just four years ago, in 1987. When OTA (1987) surveyed the public in 1986, it found overwhelming public support for field testing of genetically engineered organisms on an experimental basis. More than eight of ten Americans (82%) thought that small-scale field testing of organisms genetically altered to increase agricultural productivity or clean up environmental pollutants should be permitted. Similarly, in 1989 a survey of North Carolinians revealed that three-fourths favored such tests (Hoban, 1990). Moreover, no identifiable subset of the American population said that it opposes the environmental release of potentially beneficial organisms on an experimental basis.

A key test of this approval is the public's reaction if the release occurs in their own community. The OTA survey (1987) asked people how they would respond if their community was selected as a test site for a genetically altered organism, such as a bacterium that protects strawberries from frost, where there was no direct risk to humans and a very remote potential risk to the local environment. Here, a majority of the American public (53%) indicated they favor (strongly or somewhat) field testing this type of genetically altered organism in their own community. However, a third of the public (32%) opposed field tests under these conditions.

Interestingly, 67% of the North Carolina sample would not oppose local field tests designed to protect peaches and apples from frost. However, if they thought the tests posed a risk to the environment or human health, most of the respondents not only indicated opposition but a willingness to contact legislators and government offices to block the tests (Hoban, 1990).

When moving from the small-scale, experimental basis to the larger, commercial basis, the rate of acceptance changes dramatically. Here a majority (53%) indicated that commercial firms should not be permitted large-scale applications, even if the risks of environmental danger are judged to be very small.

The only subgroup to approve large-scale use were those who believed that dangers from genetic engineering are unlikely, and, even in this instance, it was only a 50% approval and 46% disapproval margin (OTA, 1987). While the questions changed slightly from the small-scale, experimental scenario to the large-scale, commercial example, the differences seem to be more the result of the implied risk associated with reduced control of large-scale commercial applications.

Other Impacts

Regarding genetically engineered agricultural products, the 1989 North Carolina survey revealed that the most desirable agricultural use of genetic engineering would be to produce more nutritious food (76%). In addition, over one-half (58%) of all respondents in this survey thought that the use of genetic engineering to produce frost-resistant crops would be desirable, while a slightly smaller percentage (53%) favored insect-resistant crops. There was considerably less enthusiasm for the use of genetic engineering to produce larger or faster growing stock, with only 32% of the respondents indicating that this would be desirable (Hoban, 1990).

Respondents were also asked if they would oppose a farmer using a variety of means to fight disease. Only 16% opposed the use of genetically engineered plants, while 33% opposed farmers' use of genetically engineered bacteria, and 40% opposed use of genetically engineered viruses to fight disease and pests. The greatest objection (68%) was reserved for farmers' use of stronger chemicals to control plant diseases (Hoban, 1990).

North Carolinians were asked about two specific applications of genetic engineering which have generated some controversy: herbicide-resistant crops and bovine somatotropin, a growth hormone that can increase the amount of milk dairy cows produce and increase the efficiency of milk production. Nearly half (48%) of the sample indicated herbicide-resistance would be very desirable, with only 11% reporting it would be undesirable. However, when asked how much a problem for water quality they considered the increased use of chemicals possibly accompanying herbicide-resistant crops, 92% thought it would be either a serious or moderate problem. The respondents viewed bovine somatotropin as more problematic. Forty percent believed that it would be undesirable, while only 13% thought it would be very desirable (Hoban, 1990).

Food Safety

There have been increasing public concerns about food safety. Few surveys, however, have assessed how consumers might respond to food products they believe to be genetically engineered. In one survey (Hoban, 1990), North Carolinians reported being very concerned (34%) or somewhat concerned (43%) about eating genetically engineered fruits or vegetables. They were even more concerned about eating genetically engineered meat or dairy products (45% very concerned and 37% somewhat concerned).

A group of North Carolina farmers was also asked how concerned they thought consumers would be about eating genetically engineered food. These farmers fairly accurately reflected consumer's concerns, estimating that 35% would be very concerned and 48% would be somewhat concerned about genetically engineered fruits and vegetables. They were equally accurate in representing consumer responses to genetically engineered meat and dairy products.

Social and Economic Concerns

The North Carolina survey (Hoban, 1990) also addressed the socioeconomic impacts of genetic engineering on agriculture and found widespread public concern. Most people (75%) agreed that genetic engineering will increase surpluses of agricultural commodities. Moreover, 71% agreed that genetic engineering will give larger scale farmers an unfair advantage over smaller scale farmers.

A subsample of North Carolina farmers was also asked whether genetic engineering would lead farmers to become more dependent on large corporations for their supplies. Nearly three-fourths of all farmers agreed with this statement. Finally, when they were asked if farmers should have more say in the types of products developed with biotechnology, 89% agreed that they should.

Moral Concerns

Some proponents have argued that the techniques of genetic engineering are simply an extension of classical breeding techniques. Others contend that direct genetic manipulation of plants and animals is intrinsically different from crossbreeding or cross-fertilization. The OTA survey (1987) found that the public was generally more aware of the classical techniques of plant and animal breeding than of genetic engineering (76% vs. 41%). Despite the public's differential awareness of the two technologies, they appear to hold similar views regarding the morality of the two approaches.

Among those who said they had heard of the classical techniques, one-quarter (26%) believed that crossbreeding to create hybrid plants and animals is morally wrong, while this same fraction (24%) believed that genetic engineering to create hybrid plants and animals is morally wrong. This response pattern is similar to values expressed by North Carolinians (Hoban, 1990). Approximately one-quarter thought that crossbreeding of plants and animals is morally wrong. While only 12% thought that plant genetic engineering is morally wrong, 38% believed genetic engineering of animals is morally wrong (Figure 5).

What is perhaps surprising is that a quarter or more of the public believe it is morally wrong to create hybrids, either by classical breeding techniques or by genetic engineering. Among those who believe there is a moral issue concerning the genetic manipulation of plants and animals, it appears that the

Figure 5. *Comparison of Morality of Genetic Manipulation of Plants and Animals and Classical Breeding Techniques.*

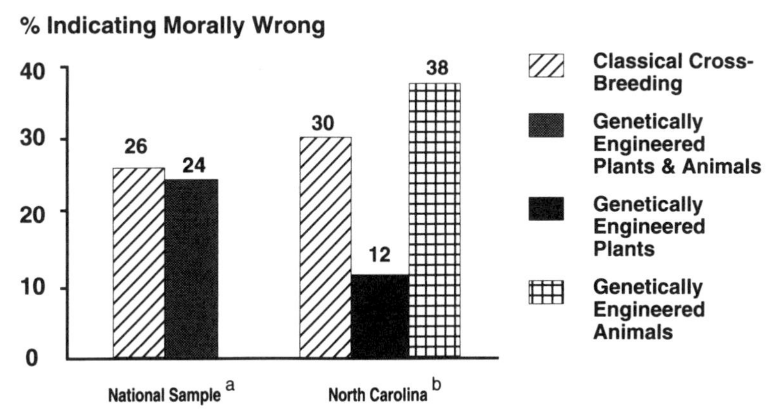

Sources: [a]*Office of Technology Assessment Survey (OTA, 1987)*
[b]*North Carolina Survey, (Hoban, 1990)*

issue lies in the objective or the fact that manipulation of any kind is occurring and not in the means by which it is achieved.

These attitudes regarding genetic manipulation were highly related to education and religiousness, with the less educated and the more religious viewing this manipulation as morally wrong. However, in the OTA survey 13% of the college graduates and 15% of those who indicate they are not religious at all reported that they feel plant and animal applications of genetic engineering are morally wrong.

Moreover, of those who feel this application of biotechnology is immoral, religious issues do not seem paramount. Just 31% of those who said it is wrong did so in terms of religious beliefs or God. Another 35%, however, objected to such applications on the grounds that "people shouldn't tamper with nature." Others expressed their concern in terms of unforeseen or unintended consequences (8%) and opposition to scientific experiments on animals (4%) (OTA, 1987). Thus, the moral objections to genetic engineering of plants and animals cover a broad range of beliefs, concerns, and fears that go well beyond religious issues.

Conclusion

The public is becoming increasingly aware and concerned about the role of biotechnology in agriculture. This concern centers on environmental impact, food safety, socioeconomic effects, and moral issues. Two-thirds of the American public see a likelihood of environmental risk from genetically altered bacteria, and nearly half see a similar risk from genetically altered plants and animals. However, given remote environmental risk and no human risk, a majority approve of the environmental use of genetically engineered organisms and field tests in their own community.

Greater concerns are expressed about food safety, with more than three-quarters of the public expressing concern about eating genetically engineered food. Similarly large percentages of the population appear worried about the negative socioeconomic consequences for farmers. Finally, while most believe application of these new technologies to plants and animals is appropriate, a quarter believe genetic manipulation is morally wrong.

Creating a Dialogue

This review of public perceptions of biotechnology reveals a diverse set of concerns over a wide range of issues. The issues are complex and include environmental and health risks, social and economic impacts, and fundamental value and moral questions. These new technologies are viewed as providing major benefits and significant costs for society and as requiring monitoring and regulation. Important questions remain regarding education and communication, as well as policy formulation for the development and introduction of genetically engineered products.

Overriding these questions is the necessity for developing meaningful dialogue and mutual respect among all parties interested in or affected by these new technologies. A number of questions must be addressed. (1) What should be the content, scope, and audience for an information and education agenda? (2) Who are the credible spokespersons to both raise and discuss the range of issues surrounding biotechnology? (3) Who should outline and articulate the broad responsibilities that come with powerful technologies like biotechnology? (4) Who should make decisions regarding the products, processes, and regulation of biotechnology? (5) How can we stimulate a productive and meaningful dialogue among the relevant constituencies?

Educational Needs

A key issue in building a more effective dialogue is to have informed and knowledgeable participants. How to ensure this is a complex issue which involves education for all the participants in the dialogue. A recent report released by the American Association for the Advancement of Science (Watkins, 1990) indicates that, if people are to understand science and its influence in their lives, science education must undergo radical reform from preschool through the university. This need has been identified as most critical at the college and university level, particularly in the undergraduate natural science curriculum.

The report maintains that today's traditional courses "often leave students with incomplete or incorrect knowledge of scientific principles, underdeveloped intellectual skills and little awareness of the influences of science on their lives" (Watkins, 1990, p. A1). To counter these outcomes, the report recommends that education in the natural sciences must be treated as a liberal art and integrated into the general education curriculum. The subject matter should be "broadened to encompass aspects of the history, philosophy, sociology, politics, and economics of science and technology" (Watkins, 1990, p. A1), and instruction should reflect the methods of practicing scientists.

The report, however, indicates that this reform of science education is as necessary for scientists as it is for the general public. The report states that by treating science as a liberal art "colleges and universities will provide society with scientists and other technically trained persons who understand science in a broad context and consequently are able to make more enlightened professional decisions" (Watkins, 1990, p. A16). The recent work on agricultural systems and ethics in public policy in agriculture by the National Agricultural and Natural Resources Curriculum Project represents a step in this direction (Wilson and Morren, 1990).

Widespread educational needs also exist for the current adult public and the scientific community. Much of the public would benefit from a brief overview of biology, as well as of how the new biotechnologies are embedded in a historical, philosophical, social, political, and economic context. Similarly, scientists and technologists would benefit from an understanding and appreciation of the broader context in which these technological developments are occurring. New organizational missions, structures, and processes may be necessary to provide this education and to promote a meaningful dialogue.

Figure 6. *Credibility of Sources of Information on Risks of Genetically Engineered Products.*

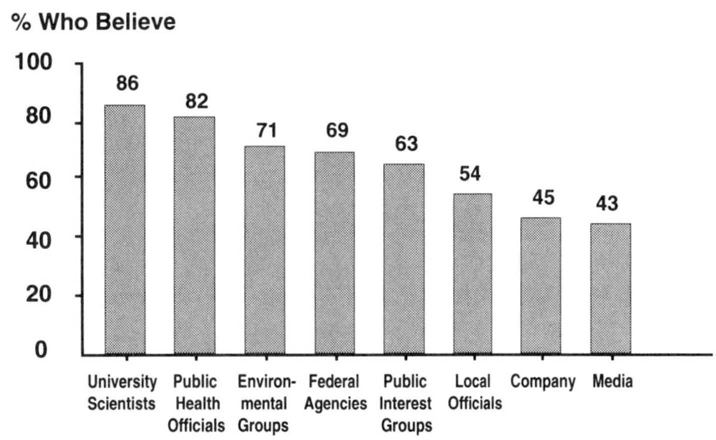

Source: Office of Technology Assessment Survey (OTA, 1987)

Credibility of Information Sources

An important factor in meaningful communication is the credibility and trust enjoyed by the various organizations and participants. Efforts to assess public perceptions of the most credible sources of information indicate that confidence and support vary considerably among the participants.

In the OTA survey (1987), the public indicated it is most likely to believe risk statements made by university scientists (86%). It also found public health officials (82%) and environmental groups (71%) to be more credible than federal agencies (69%). Still, federal agencies enjoyed more credibility than public interest groups (63%) and local officials (54%). Only a minority indicated an inclination to believe statements about risk by the company making the product (45%) or the news media (43%) (Figure 6).

These findings are nearly identical to results from a North Carolina survey (Hobañ, 1990) and a 1990 Pennsylvania survey (W. Lacy, 1990, unpublished data), where respondents indicated that they have the most faith in university scientists and the least amount of faith in the media and in the company making the product. Farmers, a subgroup of the North Carolina survey, placed greatest faith in the state Agricultural Extension Service, followed by university scientists and the U.S. Department of Agriculture.

Often, however, credible sources disagree, and the public is faced with the dilemma of choosing among expert opinions. As noted above, the OTA survey (1987) revealed that the public is inclined to believe both federal agencies (69%) and environmental groups (71%). However, these two groups may offer conflicting views on a specific topic. Thus, the public was asked which group it would believe if a federal agency reported that a genetically altered organism did not pose a significant risk but a national environmental group said it did. A significant majority (63%) indicated that they would believe the national environmental group, while only 26% would believe the federal agency.

This confidence in environmental groups is also reflected in support for their leadership. While relatively few respondents indicated that they are active in environmental groups or organizations (6%), a majority (56%) of the American public believed that on the whole the leaders of the environmental movement are responsible with their criticism and demands.

On the other hand, a majority (57%) of the American people believed that the leaders of the environmental movement are out of touch with the public, while only a third (35%) said that the leaders of the environmental movement reflect public feeling (OTA, 1987). Thus, the public tends to view environmental groups and their spokespersons as presenting valid criticism and reasonable demands, while it may not necessarily subscribe to the entire value structure of these advocates.

In summary, the public appears to place greatest confidence in information from university scientists. However, confidence in those scientists may be decreasing, as indicated earlier by the decline in confidence in science and technology over the last three decades. In addition, the public views environmental groups as highly credible sources of information which it would believe over federal agencies. The least amount of confidence resides in the media and the very companies and corporations who are developing the new products and processes.

Who Should Decide?

An important question remains regarding who should decide upon the products and processes to be developed and the specific regulations to govern commercial applications. As noted previously, the public overwhelmingly believes that strict regulation of technology is necessary because it perceives potential dangers from these innovations.

In the OTA survey (1987), respondents were asked who should decide whether or not commercial firms should be permitted to apply genetically altered organisms on a large-scale basis: the company that developed the product; an external scientific body; a government agency; an industrial trade association; or other group. A government agency is the most frequently cited source for defining commercial applications (37%), with an external scientific body preferred by 29%. Only 13% felt that the decision should be left to the company which developed the product. The public, voters, and the community, categories not included in the questionnaire, were volunteered by 5% of the respondents.

In the Nova Laboratories survey (Russell, et al., 1987) the public demonstrated its trust in the scientific community by suggesting that scientists be involved in the regulatory process. Among those groups most frequently cited for participating in the establishment of specific regulatory controls are the American Medical Association (70%), the scientists who conduct genetic engineering research (68%), the federal government (62%), and the people concerned with social ethics (60%).

A group of experts interviewed as part of the Nova Laboratories survey concurred with the suggested public involvement in the regulatory process. A majority of these experts favored regulation developed by a broad coalition of people from different disciplines, including scientists, ethicists, and lay people. They noted that such a coalition requires an informed public.

In this same survey of the public, two-thirds (66%) favored an international agreement establishing a common set of guidelines and limitations for company scientists involved in genetic engineering worldwide. Those most frequently mentioned to provide the necessary monitoring were an international group of scientists (77%) and the United Nations (65%) (Russell, et al., 1987).

In the North Carolina study, over four-fifths (81%) indicated that the states should have their own regulations. Moreover, only half (52%) felt that the federal regulations are adequate. Finally, most of the respondents (87%) believed that the public should have more say in government regulation of genetic engineering (Hoban, 1990).

Stimulating the Dialogue

In conclusion, it is clear that these perceptions and public opinions about biotechnology research and development are framed by the public in terms of value questions in a democratic society. Efforts should be made to consider a broad range of issues, including environmental and health risks, the social and economic impacts, and fundamental value and moral questions.

Moreover, the science and technology educational needs for all segments of the population should undergo a major reexamination and reform. Furthermore, the dialogue should involve a broad base of participants and organizations, including ethicists, philosophers, religious leaders, farmers, environmentalists, scientists, the media, professional societies, businesses, governments, and educational institutions.

The major goal in establishing this dialogue should be to generate communication which addresses the concerns of all legitimate parties, rather than targeting the limited needs of a single client. This communication must not be limited to educating people in the use of probabilities or to resolving the conflicts which inevitably arise in society over the implementation of technologies. Further, meaningful communication must avoid characterizing the public as naive, irrational, illogical, and scientifically illiterate. Instead, it must provide the context for an open forum and public debate from which both scientists and the public can learn.

These forums and public debates must be combined with a careful and open examination of the various concerns raised by the public regarding the development and commercialization of biotechnology. They must also go beyond the present,when possible, to anticipate future consequences, drawing upon the latest knowledge so that the right questions can be asked.

Federal and state governments, universities, and private foundations have important roles to play in support of the public debate about science policy and biotechnology. There are some positive indications that this is already occurring.

One institution, the Keystone Center, a national, private organization dedicated to facilitation, mediation, and education, was founded in 1975 to promote consensus-building approaches to complex and controversial science-related issues at the local, state, national, and international levels. Through funding by the Joyce Foundation, the Keystone Center established the Environmental, Citizen, State and Local Leadership Initiative for Biotechnology in 1987.

The objective of this program is to disseminate information and to encourage discussion among environmental, citizen, and state and local government leaders regarding public policy issues associated with the development of biotechnology. They have conducted regional workshops and published several reports to promote effective participation in the emerging policy debate surrounding biotechnology (Keystone, 1989).

A university structure to promote dialogue was established in 1985 by the University of California Systemwide Biotechnology Research and Education Program. One of its major purposes was to inform those in government, industry, and the public about biotechnology developments and the implications of these developments in the public arena. The program has implemented a call for proposals for (1) critical study of the regulatory policies, risk/benefit assessments, private and government economic support, and social and ethical implications of biotechnology and (2) training of personnel who can help industry, government, and society understand and respond to arising issues. Current support permits initiation of up to four awards in 1989 and 1990, each with a possible commitment of up to $100,000.

A third organization, The National Agricultural Biotechnology Council, more specifically targets agriculture. A consortium of the Boyce Thompson Institute for Plant Research, Cornell University, Iowa State University, the Texas A & M University System, the University of California at Davis, and Purdue University, it was created in 1988 with support from the Joyce Foundation and the U.S. Department of Agriculture. It provides a forum for the scientific community, agribusiness, farm groups, and government, environmental, and consumer groups to analyze and debate the potential impacts of agricultural biotechnology and to develop recommendations and policy options. In the spring of 1990, the council held its second annual meeting,"Agricultural Biotechnology, Food Safety and Nutritional Quality for the Consumer."

The council cosponsored the meeting with the Agricultural Research Institute, founded in 1951 by the Board on Agriculture of the National Academy of Sciences. The latter nonprofit consortium of public and private organizations also analyzes and publicizes problem areas in agriculture and serves as a forum where public and private officials can exchange views and generate actions.

The Cooperative Extension System is a key organization capable of promoting a meaningful dialogue on agricultural biotechnology. This community- and university-based educational system is uniquely positioned to provide leadership in this area, since it links research to formal and informal community education, has both a practical orientation and experience with a wide range of audiences, and enjoys credibility with rural and city residents, as well as local, state and federal officials.

To strengthen the capacity of this system to stimulate and organize effective public policy discussions on biotechnology, the Extension Biotechnology Coordinator Network has been recently established. This network includes scientists and extension leaders from each of the land-grant universities and serves as an important linkage between state extension programs, relevant federal agencies, and biotechnology firms in the private sector (Weber, 1990).

In the future, these organizations and others must continue to bring together the most diverse and best thinkers from local, state, and national governments, and from education, business, and public interest groups to debate, to work together, and to engage in an effective dialogue. This dialogue must ensure democratic participation in decision making about agricultural biotechnology.

It should seek to empower the diverse elements of the public so they are confident that their views about biotechnology will not only be heard, but make a difference in the decision making. The goals of this dialogue should not be public acceptance but public understanding, not the application of biotechnology but developing responsible decisions regarding biotechnology applications which are compatible with community goals and values.

Finally, the process of open debate and meaningful dialogue is not without personal and institutional costs. Congressman George Brown reminds us that we must be willing to live with the outcomes of this dialogue, even if they run counter to our wishes and desires. Only by participating in the process with this attitude will it be worth the effort. A New York Times editorial recently noted, that ultimately "the rates and risks of progress are matters for society to decide" (as quoted in Russell, et. al., 1987). As recipients of public funds, the public agricultural research community has an opportunity and an obligation to assist society in making such decisions in a democratic and informed manner.

The Cooperative Extension Service is a key organization in promoting a meaningful dialogue on agricultural biotechnology.

References and Additional Readings

Berrier, R. J. 1987. Public perceptions of biotechnology. p. 37-52. In *L. R. Batra and W. Klassen (eds.) Public perceptions of biotechnology. Agricultural Research Institute, Bethesda, MD.*

Brown, G. E., Jr. 1987. Social and political aspects of risk assessment. p. 73-80. In *L. R. Batra and W. Klassen (ed.) Public perceptions of biotechnology. Agricultural Research Institute, Bethesda, MD.*

Busch, L., W. B. Lacy, J. Burkhardt, and L. R. Lacy. 1990. Plants, power and profit: Social, economic and ethical consequences of the new biotechnologies. Basil Blackwell, London, England.

Cannell, W. and H. Otway. 1988. Audience perspectives in the communication of technological risks. Futures 519-531.

Hoban, T. J. 1990. Agricultural biotechnology: Public attitudes and educational needs. North Carolina State University, Raleigh, NC.

Kalter, R. J. and L. W. Tauer. 1987. Potential economic impacts of agricultural biotechnology. American Journal of Agricultural Economics 69:420-425.

Keystone Center. 1989. Biotechnology decision making: Perspectives on the objectives of public participation. The Keystone Center, Keystone, CO.

Klassen, W. 1989. Public concerns regarding applications of biotechnologies. Biodeterioration Research 2:7-17.

Lacy, W. B., L. R. Lacy, and L. Busch. 1988. Agricultural biotechnology research: Practices, consequences, and policy recommendations. Agriculture and Human Values 5(3): 3-14.

Mellon, M. 1987. Biotechnology and the environment: A primer on the environmental implications of genetic engineering. National Wildlife Federation, Washington, DC.

Office of Technology Assessment, U.S. Congress. 1987. New developments in biotechnology: Public perceptions of biotechnology. OTA-BP-BA-45. U.S. Government Printing Office, Washington, DC.

Russell, V., R. Denny, S. Parnett, F. Walton, D. Merriam, and K. Barnes. 1987. The Nova Report: American attitudes and beliefs about genetic engineering. Research and Forecasts, Inc., New York, NY.

Sagoff, M. 1988. Biotechnology and the environment: What is at risk? Agriculture and Human Values 5(32): 26-35.

Sandman, P. M. 1987. Apathy versus hysteria: Public perception of risk. p. 219-232. In *L. R. Batra and W. Klassen (eds.) Public perceptions of biotechnology. Agricultural Research Institute, Bethesda, MD.*

Watkins, B. T. 1990. Science education seen in dire need of radical reform. The Chronicle of Higher Education 36(34): A1, A16.

Weber, G. 1990. USDA-ES. Extension biotechnology coordinators memo. Washington, DC: March 24:1-2.

Wilson, K. and G. E. B. Morren, Jr. 1990. Systems approaches for improvement in agriculture and resource management. Macmillan Publishing Co., New York, NY.

CHAPTER ELEVEN

This last chapter serves as a capstone to the book. It addresses several of the critical issues and choices suggested in previous chapters. In addition, the chapter examines some of the positive and negative impacts that research and development in biotechnology may have on agriculture and related industries.

The chapter analyzes the future impact of agricultural biotechnology by defining a set of issues and choices. These choices will be made by those who: (1) do research in biotechnology, (2) make science policy and provide funding for research and development, (3) are involved in education of scientists and the general populace, (4) make investments and operational decisions in industry, (5) establish and implement regulatory policy on biotechnology safety, and (6) consume the food and fiber that come from our broadly defined food and agriculture system.

11 Issues and Choices: How Will Agricultural Biotechnology Influence the Future?

Neville P. Clarke, Texas A&M University
Charles E. Hess, Science & Education-USDA

Introduction

The tools of biotechnology and the new knowledge and products made possible by using them will be vitally important to the food and agriculture system. Moreover, the knowledge and tools come at a time when we are dealing with challenges affecting the entire food and agriculture system throughout the world.

These larger challenges involve overarching issues such as (1) the need to make food available at affordable prices for a burgeoning population; (2) the recognition of the importance of natural resource stewardship, including sustainable production systems and the impact of global climate change; (3) the internationalization of agriculture as part of the continuing emergence of the global village that links all peoples, especially in the area of food and fiber; and (4) the growing concern about the health and safety of what is consumed and about the impact of the by-products of production on human wellness in the broadest sense.

Most scientists believe biotechnology has the potential to positively affect how we meet most, if not all, of these larger challenges. However, scientists, policy makers, and consumers also are concerned about the possible risk involved in emerging biotechnology applications, notably the risk of negatively affecting the environment and human health. Some also wish to see the socioeconomic consequences of biotechnology applications addressed before, not after, they are applied, (i.e., *ex ante* analysis).

Clearly, improved communication between scientists, those concerned with the environment and food safety, and the "ultimate decision maker," the consumer, must be developed as part of the emerging agenda. Perhaps more than any single variable, the success of this communication will have a profound impact on the emergence and use of biotechnology in the years ahead. To be effective, this communication must be multi-, not unidirectional. It is our hope that this chapter will foster that communication.

Critical Issues and Choices

The authors of the previous chapters have examined the expected impacts of biotechnology and have either implicitly or explicitly posed a number of issues that help define many of the choices and options which are emerging as biotechnology research and development continues and is transferred into use and practice. The following examines a number of these critical issues and offers some choices.

Issue: *Where will biotechnology research be conducted?*

Choices: *Conduct biotechnology research in the public sector, or conduct it in the private sector.*

Most knowledgeable people believe that the advanced research in plant molecular biology is now occurring more in private industry rather than in academia. To a lesser degree, this also may be the case with animal agriculture and microbiologic research. This has occurred because of the perceived opportunity to commercialize the products of biotechnology, which has resulted in rapid investments in facilities and equipment and in the recruitment of many of the best basic scientists into industrial laboratories.

In the early 1980s there was a substantial buildup of basic research in biotechnology industries. Despite a major shakeout of venture-capital-backed industries in the early to mid-1980s, today a substantial amount of that type of research is still being done there. As biotechnology research and development activities in industry have matured, however, there is a growing endorsement by industry of the need for academia to do the high-risk, cutting-edge research. Thus, while both academia and private industries conduct basic and applied research, the emphasis in academia is basic research, with private industries focusing on more applied research.

If there is a resurgence or growth of biotechnology research in academia that is related to the food and agriculture system, will it occur in the land-grant institutions or non-land-grant institutions? Perhaps a more direct question is, What should be the balance between these two classes of institutions?

A recent Office of Technology Assessment report states that 35 percent of federally sponsored research directly affecting agriculture is done outside of USDA. Medical research in biotechnology has contributed significantly to animal agriculture research and continues to provide many of the fundamental breakthroughs in methodology.

Many departments of basic biology and related science in academic institutions that are outside the traditional colleges of agriculture are making very substantial contributions in both basic and applied biology related to agriculture. There are increasing numbers of broad linkages among various entities in academia and between academia and industry.

Many scientists and funding agencies feel that academia must be involved in the basic research agenda if it is to provide the future generations of scientists and technicians that will be needed to meet expanding staff needs in industry and academia. Also, industry, by its nature, focuses on relatively short-term, market-driven research. Thus, there is a need to provide through the public sector a stable, science-driven research foundation for agricultural industry.

There is a need for universities to provide a stable, science-driven research foundation for agriculture.

The critical choices involve the level of biotechnology funding, its sources, and its allocation among private firms and academia. Scientists, administrators, users of technology, and taxpayers all have a role in these decisions.

Issue: *Should biotechnology research be problem driven or science driven?*

Choices: *Continue federal government funding of obvious, short-term applications, or provide funding for science-driven research.*

Neither choice is desirable to the exclusion of the other. But there is a clear lack of appropriate balance between basic and applied research in the life, engineering, and social sciences that support the food and agriculture system. The federal government funds basic research in the USDA's Agricultural Research Service (ARS) and provides Hatch support for basic and applied research in the land-grant experiment stations.

Less than five percent of the USDA's total research budget is spent on competitive grants in agriculture. Thus, the National Initiative for Research on Agriculture, Food and Environment is currently advocating new funding at the level of $500 million per year, with emphasis on acquisition of fundamentally new knowledge and technology for agriculture. The initiative would, if fully funded, shift the competitive grants investment to about 25 percent of the total federal funding effort.

***Issue:** Is the use of chemical inputs in agriculture excessive?*

***Choices:** Have U.S. farmers continue to rely on chemical inputs, or replace chemical inputs with methods and products derived from biotechnology.*

Restrictions placed on the chemicals available for agriculture are expected to continue to increase whenever these products have, or are perceived to have, a potential negative effect either on the environment or human health. Development of alternative chemicals that are efficacious, affordable, and either environmentally neutral or positive is one approach. Another approach is to emphasize biological alternatives to the use of chemicals. Combinations of these general methods will likely be employed in the next decade or two.

***Issue:** Who will be the beneficiaries of agricultural biotechnology?*

***Choices:** Private firms that develop the technologies or technology users or consumers will be the major beneficiaries.*

A possible impact of biotechnology research, as stated earlier, could be that the technology gets locked up in a few large, vertically integrated industries. What are the conditions under which this could happen? Are there other alternatives?

As industry makes use of biotechnology or any other kind of technology, it usually does one of three things: (1) it protects its position with patents and licenses (which require disclosure of the technology); (2) it develops and keeps secret certain know-how that allows it to produce a unique product; or (3) it gets a sufficient lead on competing industry to maintain an attractive market share, whether or not its technology is protected.

While a new plant or animal variety might not be released by a large, vertically integrated company, particularly if it provided a unique or unusual market quality, there are many other kinds of technology employing recombinant DNA methods that would find their maximum return on investment through large volume sales to others involved in the food and agriculture system. Some technology (nonhybrid seed, for instance), once released by the manufacturer, becomes difficult to protect and may be further used without dealing with the parent company.

Many of the projected products of biotechnology research are "size neutral," that is, equally affordable and usable by small and large agricultural industries. Clearly, the volume of product sold will be related to the perceived cost versus gain to the purchaser. For many of the biotechnology-related industries, this market principle will provide insurance that new products will be available at an appropriate price.

Some products from biotechnology will require more intensive or skillful management, which many believe will create a less-than-level playing field for the smaller (part-time) farmer. In cases where new technology may be size dependent, existing systems such as the Cooperative Extension Service may be useful in mitigating adverse impacts where possible.

Barring a technology lockup by a small segment of industry on a critical facet of production or manufacturing, it would appear that the general case regarding the beneficiary relationship between producer and consumer would hold for new products developed from biotechnology. Early innovators in industry (or those who can maintain a privileged use) would benefit most from new technology.

Ultimately, consumers should benefit most either through lower cost of food and fiber due to efficiency of production or through advantages related to quality, safety, or convenience. Market forces, consumer preferences, and government policies will determine the distribution of the benefits from agricultural biotechnology.

***Issue:** How should agricultural biotechnology be regulated to ensure biosafety?*

***Choices:** Implement voluntary guidelines, or establish mandatory requirements for scientists.*

The Recombinant DNA Advisory Committee (RAC) of the National Institutes of Health (NIH) has provided a methodology for contained research that basically decentralizes decision making on biosafety to Institutional Biosafety Committees (IBCs) in academia and industry. The deregulation of responsibility to the IBCs increases as experience provides confidence for certain types of research.

However, some who are concerned about environmental release of living products of biotechnology believe that stringent regulation of such releases must be maintained. The issue is how best to employ a combination of continuing regulatory oversight and local decision making for biosafety in biotechnology research. Both are needed, but avoiding unnecessary regulation is essential for productivity of research and its application.

The issue is how best to employ a combination of continuing regulatory oversight and local decision making for biosafety in biotechnology research.

IBCs in universities and industry have responsibility for safety in all research using recombinant DNA methods. Some procedures require the concurrence of the IBC only; others require additional approval by appropriate federal or other regulatory authority. Guidelines for biosafety in agricultural biotechnology are being released. Regulatory authorities in the Environmental Protection Agency (EPA), USDA, Federal Drug Administration (FDA), and other agencies are operating within existing legislative authority by expanding or modifying regulations to deal with the emerging needs in biotechnology.

The development of workable and understandable regulatory mechanisms has been slow to occur, but there has been progress in recent months. There is an urgent need to enunciate a clear and coordinated policy on federal regulation of biotechnology research that is consistent across agencies and avoids proliferation of local and state regulations that often confound and confuse.

Many state and local regulatory authorities either have or are now developing separate requirements for conducting research and development as well as for release and use of products from biotechnology. Clear federal processes and procedures must emerge to avoid a plethora of individual, confusing, and perhaps overly constraining state or local limitations on biotechnology.

Because food and other agricultural products must be able to enter the international marketplace freely, it is obviously critical that international standards and procedures are in place to assure an agreed-upon methodology for controlling and ensuring the safety of the products of biotechnology. The USDA and other governmental agencies are actively involved in developing these international standards at the present time.

As experience and practice allow, there should be a continuing effort to relax the regulatory process in areas where experience demonstrates an acceptably low level of risk. The ideal situation will be when the Institutional Biosafety Committees (IBCs) can approve new research based on solid previous experience demonstrating safety, without burdensome and lengthy approval by regulatory authorities. By the same token, review and approval must be maintained where lack of experience and real or potential danger exists.

There is substantial concern—and growing evidence—that unless workable regulatory processes for development and testing of products derived using biotechnology methods are agreed to and put in place, the next generation of biotechnology products either will not be developed or will be developed in countries other than the United States. A number of highly capable scientists and faculty in academia are simply putting early results on the shelf. They are unwilling to deal with the uncertainties associated with the present regulatory processes.

The USDA has a new program in place, the National Biological Impact Assessment Program (NBIAP), that offers assistance and information to investigators or institutions contemplating biotechnology research. Data of various types, assistance in regulatory processing, and information on recommended experimental design and available test sites are readily accessible through automated data bases and expert system-based technology. These services are intended to facilitate compliance with existing regulations by easing the burden on individual scientists.

Issue: *What should be the criteria for exposure to hazardous material in food?*

Choices: *Maintain zero-exposure criteria for certain classes of hazards, or develop methods of risk assessment that expand the concept of reasonable exposure.*

It is vitally important that a new look be taken at the general subject of risk assessment relative to food safety and wholesomeness. New biological data are obviously needed on safety and wholesomeness. But there also is need for improved methods of assessing toxicity and carcinogenicity and of statistically expressing consequences other than zero-tolerance limits.

The Hazard Assessment At Critical Control Points methodology offers an important methodology applicable to products of biotechnology. This concept should be expanded further and linked to a more general statistical model of risk.

Assessing risk and informing technology users and the public are essential. Scientists must not only offer accurate statistical assessments of risks but also associate any risk with the potential benefits. Such information can help the general public make more informed judgments on their willingness to encounter some risk while enjoying benefits from a particular application of biotechnology.

Issue: *What will be the impact of biotechnology on natural resources?*

Choices: *Accept very low risk for high benefit, or accept no environmental risk.*

Many public concerns about the potential negative impact of biotechnology center on its effects on natural resources and the environment. However, there also are compelling arguments suggesting that the positive effects will be very attractive to those concerned about the prudent use and protection of natural resources. For example, bioremediation offers the opportunity to use genetically engineered organisms to degrade toxic substances in the environment.

In the area of biological control, the use of recombinant DNA methodologies offers a major opportunity to produce biological alternatives to chemical pesticides and fertilizers. A new generation of biological control research is emerging and gives great promise for expanded use of integrated pest management (IPM). The prospects are excellent for broadening the use of this and related management methods to a greater range of crops through biotechnology.

The new alternative agriculture practices related to sustainability call for revitalization of older methods and development of new techniques in farming operations that can reduce the use of chemical inputs through management and use of natural resources in lieu of purchased inputs. The National Research Council's report on this subject lays out a set of options and practices that lend themselves well to enhanced performance through the use of biotechnology to improve performance in areas such as nitrogen fixation, pest resistance, and allelopathy.

Ecologists are expressing growing concerns about loss of biodiversity, especially in wilderness areas, but also in developed areas of increasing intensity of agricultural operations. Biotechnology offers new opportunities to preserve and maintain endangered species, both animal and plant, and to provide reservoirs of genetic material for future expansion. A current example of this use lies in the preservation of exotic and endangered germplasm of wild animal species through embryo transfer and preservation.

Earlier chapters have highlighted the concerns and the precautions that must be recognized to assure that there is not a "biological accident" resulting from the use of recombinant DNA methods in agriculture and the environment. Varying levels of potential danger have been recognized, and corresponding levels of regulatory and practical constraints need to be continuously defined to minimize the potential for such accidents.

Plant and animal breeders, who have decades of practical experience with development of new germplasm, believe that the precision of genetic manipulation using recombinant methods is greater, and, therefore, the result is more predictable than with traditional methods. The general strategy for risk management is to employ a conservative approach at the outset and to relax constraints on experimental procedures as practice and experience can justify. This is similar to the approach that has been used over the last 15 years in dealing with recombinant DNA research in contained environments.

Clearly, the complexity of natural ecosystems and the lack of full understanding of the factors that can have an impact on such systems makes environmental release of genetically modified organisms correspondingly more complex. But we are already gaining a positive experience with such releases, amply demonstrating that reasoned planning and regulatory oversight can provide the assurance needed to use this methodology effectively and safely.

Issue: *What will be the impact of biotechnology on the international agricultural marketplace?*

Choices: *Use biotechnology to develop agriculture in low-income countries, or allow the United States to become more competitive in international markets.*

All food and agriculture industries operate either directly or indirectly in the international marketplace. There are established and expanding international standards and practices that are employed to assure the safety and wholesomeness of food and the quality of nonfood agricultural products.

Recent efforts have been made to apply those standards and practices pertaining to international safety regulations and trade agreements to the products of biotechnology. In addition to dealing with the products of biotechnology, such safety considerations are applicable in principle to the research methods used to produce genetically engineered products.

While some standards and practices are still emerging, several general principles have been agreed to: (1) general standards of safety should be uniformly applied among countries; (2) exported products of biotechnology must be produced under safety guidelines comparable to those of the importing country; (3) general definitions and terms of understanding must be developed in the international community to assure the ability for explicit communication; (4) methods for assuring flexibility in modifying international standards based on experimental and production experience are mandatory; (5) safety considerations with products of biotechnology should not be used as artificial trade barriers; and (6) socioeconomic impacts of products and procedures using biotechnology should neither be treated in any way that is conceptually different from other new technology, nor should they be incorporated into a regulatory framework.

With reference to the public support of research, concern has been expressed about the export of biotechnology science and "know-how." Farmers and their representative organizations feel that U.S. agriculture should have preferential use of the new technology developed in the United States. However, much of the investment in agricultural biotechnology is being made by multinational companies; and many of these companies see, at least in the short-run, their most exciting prospects for commercialization in other countries.

Previously, it has been assumed that early users of new technology would be in a preferred position. This has certainly been true in nonagricultural industries. The extent to which this will remain would appear to be affected by the extraordinary rapidity with which new biotechnology is transferred from the laboratory to the marketplace.

While the time scale may be different, the general picture for agricultural biotechnology is likely to be the same as for industry in general. It is not, for instance, reasonable to expect that a national technology lockup can occur for long periods in any country that will effectively protect its competitive position in the international marketplace. Also, recognizing the site specificity of agricultural technology, products of biotechnology will often have maximum utility in geographic areas or production systems for which they are specifically created.

Early users of new technology will continue to have an advantage. And, as with other industries, the global food and agriculture system is on a "technology treadmill." In other words, it is likely that existing technology will be replaced by new technologies, and technological competition in industries that operate in the international marketplace will continue.

Issue: *How will biotechnology affect the structure of agriculture?*

Choices: *Continue or accelerate the growth of large operators, or create size-neutral technologies.*

There is excellent treatment of this complex subject in earlier chapters. The major agricultural industries that have provided farmers and ranchers with chemical inputs have already made enormous investments in biotechnology. This is in recognition of the major opportunity for new, profitable products from biotechnology and the inevitable reduction in the use of environmentally threatening chemicals in agriculture.

Production agriculture will be affected by biotechnology as it influences the input industries, the processing and marketing industries, and the international marketplace. Production agriculture is the focal point for the application of biotechnology and for the concerns about the biological and social consequences of such applications.

Post-harvest processing and manufacturing of food and fiber products offer major opportunities for expanding overall economic activity of the food and agriculture system. Using biotechnology to create new products, improve existing products, and develop new methods of quality and safety assurance broadly across the food production system are all exciting new prospects.

Biotechnology offers a particularly exciting opportunity to reduce hazards related to safety and quality of existing food. The recent tragedy of the marketing of a tryptophan product found to be highly toxic highlights the importance of thorough and continuing quality and safety assurance procedures in the processing and manufacturing industries, regardless of the kind of technologies used in development and manufacture.

Separate and apart from the applications of biotechnology, a progressive and substantial growth of vertically integrated, multinational industries is occurring. As indicated earlier, such industries have a comparative advantage in using the products of biotechnology. There are many products of biotechnology which, to be profitable, must develop large markets. For such products to be successful, they must provide a desired capability, and they must be available at prices that make economic sense to the purchaser.

There may be emerging at least one dimension of a "division of labor" between industrially and publicly supported research in terms of the ability to protect the resulting intellectual property. Public institutions might focus on development of new knowledge and technology that is broadly applicable and difficult to patent or otherwise protect, while industry is already focusing on those products that are intended to provide a profit in the marketplace.

Production agriculture is the focal point for the application of biotechnology and for concerns about the biological and social consequences of such applications.

Issue: *How should biotechnology products be marketed?*

Choices: *Develop national programs to foster the commercial applications of biotechnology discoveries, or continue exporting new technology for foreign exploitation.*

Other countries, as exemplified by Japan, have an excellent track record in taking early technology from the test tube to the pilot plant in the shortest possible time. The reason for this, in many cases, is that the large "front-end" investment in basic research in the United States is exploited by other countries which seem to be better able to organize and execute the product-development process.

While it seems fair to say that university-industry relationships (including early partnering) have become commonplace in the decade of the 1980s, there is still a need to optimize this process in the United States if we are to avoid losing the benefits of basic research. The changes in patent laws in the early 1980s and the Federal Technology Transfer Act of 1986 have provided valuable increased incentives for early use of new technology. Methods to encourage entrepreneurship, the removal of unnecessary regulatory barriers, the making of capital more readily available for early ventures, and providing public funding for the early development phase of research in biotechnology are all methods that have been suggested as partial solutions to this dilemma.

Issue: *Should the* ex ante *estimation of socioeconomic consequences of research be part of the biotechnology approval process?*

Choices: *Continue present scientific criteria to evaluate new technology, or include* ex ante *analysis of the consequences of research as a part of the planning, funding, and approval process.*

Methods for performing *ex ante* evaluation of the socioeconomic impact of research are imprecise but improving. Some policy makers are calling for an earlier assessment of the impact of research on farmers, rural communities, consumers, and governments. The implication is that, should negative socioeconomic consequences be forecast, funding for certain research would be withheld and/or those biotechnology products would not be approved by government agencies.

Whether this analysis is done at the individual project level or for classes of research is part of the consideration. One early version of the 1990 Farm Bill called for *ex ante* assessment of research to assure its applicability to certain classes of farmers and agricultural operations.

Those involved in biological and engineering research have been charged with being insensitive to the social and economic impacts of the research they do. It is fair to say that scientists from these disciplines often are not totally aware of such consequences or how they might be assessed. They, like the users of their technology, have not been motivated in the past to consider such concerns.

Social and economic impacts of technology, including biotechnology, are valid and valuable areas of research, and those findings should be taken into consideration—at the appropriate stage. They could act as artificial constraints in early-stage research, and thus stifle the scientific creativity that is the source of new technology. The findings of socioeconomic impact research are more appropriate at the applied-research and technology-adaptation stages, when there is actually something with potential impacts to be assessed.

Issue: *How should consumer attitudes about the products of biotechnology be addressed?*

Choices: *Address consumer concerns about specific products in the marketplace as they arise, or undertake a broad initiative to expand the biotechnology dialogue.*

Consumers are being made more conscious of environmental issues in schools, by the news media, and in general literature. Similarly, a growing emphasis is being placed on diet-health issues and food safety. There is a tendency to attach these types of concerns to the processes and products related to biotechnology.

Because bad news tends to travel faster and further than good news and because sensationalism has been successfully implemented against some products of biotechnology, a proactive educational initiative, rather than a reactive, product-specific response, seems appropriate. The public has a right to and a need for factual, credible information about biotechnology.

Interactions between social and biological issues are complex, dynamic, and profound. Recent conferences on biotechnology, especially as it pertains to food safety and wholesomeness, have placed consumer communication at the top of the priority list. While there is no question of the importance of this need, there is still a lack of a definitive and workable plan for providing information and opportunities for public dialogue.

The traditional sources of information on food safety have a credibility problem with many consumers. A substantial erosion of consumer confidence has been repeatedly measured in the last two years. This has led to consideration of the possible need for a separate entity, uncoupled from both government and industry, representing the diverse food safety concerns; a sort of "Underwriters" group to evaluate and develop "labeling" information about food safety. This fits into a much broader agenda of food labeling that is emerging from the consumer-driven organizations that work with and in the government.

Consumers' concerns about food safety are obviously much broader than their concerns about products of biotechnology in food. A wholesome, healthy diet is a national priority that has been well established in the minds of many consumers. This subject must be part of a broader educational effort. For some consumers, food products developed through biotechnology are not being perceived as part of the solution, but as part of the problem in assuring wholesomeness and safety.

The science community and the regulatory community in government contend that the products of biotechnology are not inherently any more or less dangerous than products developed using traditional methods. There is a growing body of practical experience to support this position.

An appropriate set of standards for assuring the safety of research is needed, especially when environmental releases of genetically modified organisms are known to be safe. But, given that this need is met and products become available in the marketplace, there is a clear and urgent need to provide a much more effective mechanism for interaction and communication with the general public than is in place today.

Conclusions

Biotechnology has made enormous progress in the last 15 years. The fundamental opportunities for biotechnology can now be seen more clearly, although new vistas are opening at every juncture. Social, economic, and biologic concerns about the impacts of biotechnology have been examined throughout this book. The need for enhanced communication and public dialogue on these issues is paramount.

The potential for major positive benefits from biotechnology is substantial. However, breakthroughs in biotechnology research and development will be required if agriculture is to meet the demands of the next century. Moreover, risks and possible negative impacts also must be carefully analyzed and assessed. Together, we all must seek to define the critical issues and make informed choices. It is to be hoped that this book has helped further this process.

Glossary

Agrobacterium A bacterium normally responsible for production of crown gall disease in a variety of plants. A plasmid has been isolated from this bacterium that is useful in plant genetic engineering. This plasmid, called the Ti plasmid, has been modified so that it does not cause disease but can carry foreign DNA into susceptible plant cells.

Anthropogenic Relating to or involving the impact of man on nature.

Antibody A protein produced by humans and higher animals in response to the presence of a specific antigen.

Antigen A substance that, when introduced into the body, induces an immune response by a specific antibody.

Bacteriophage A virus that lives in and kills bacteria.

Bioreactor A vessel or container in which a biological reaction is performed.

BST Bovine somatotropin. The version of somatotropin produced by cattle.

Bt *Bacillus thuringiensis.* A bacterium that produces a toxin which is lethal to certain insect larvae. This bacterium, in killed form, has been used for about three decades as a biological control agent. Much research has been conducted to insert the gene that produces the *Bt*-toxin into plants in order to make the plants insect resistant.

Callus A cluster of undifferentiated plant cells which can develop into a whole organism (complete plant).

Chromosome Threadlike components in the cell that contain DNA and proteins. Genes are carried on the chromosomes.

DNA Deoxyribonucleic acid. The molecule that carries the genetic information for most living systems.

DNA probe A short section of single-stranded DNA designed to be complementary to some characteristic part of the DNA to be tested. Probes are labeled, with radioactivity, for example, so that their binding can be easily detected.

E. coli *Escherichia coli.* A bacterium that inhabits the intestinal tract of most vertebrates. Much of the work using recombinant DNA techniques has been carried out with this organism because it has been genetically well characterized.

Edaphic Resulting from or influenced by the soil rather than the climate.

Enzyme A protein catalyst that facilitates specific chemical or metabolic reactions necessary for cell growth and reproduction.

Epidemiology A science that deals with the incidence, distribution, and control of disease in a population.

Explant Living plant tissue that has been removed from a plant into tissue culture.

Fermentation The conversion of one substance into a more desirable substance through the actions of microorganisms.

Gene A segment of chromosome. Some genes direct the syntheses of proteins, while others have regulatory functions.

Gene mapping Determination of the relative locations of genes on a chromosome.

Genetic promoter A DNA sequence that is located in front of a gene and controls gene expression.

Genome The complete genetic code for any individual, the genetic sum of its DNA.

Genotype The genetic makeup of a plant or animal that gives rise to its visible characteristics.

Germplasm The total genetic variability, represented by germ cells (reproductive cells), available to a particular population of organisms.

Hybrid The offspring of genetically dissimilar parents, whether cells, plants, or animals.

Hybrid DNA DNA produced by combining new or foreign DNA with host DNA at the molecular level.

Hybridization 1. Production of hybrid offspring. The process can be used to produce hybrid plants or hybridomas. 2. Binding of complementary strands of DNA or RNA.

Hybridoma The hybrid cell produced by fusing two cells of different origin. In monoclonal antibody technology, hybridomas are formed by fusing a cancer cell and an antibody-producing cell.

Immunoassay Technique for identifying substances based on the use of antibodies.

Metabolites Substances that are substrates for or are produced by metabolic processes or enzyme reactions.

Appreciation is expressed to the Industrial Biotechnology Association for permitting the use of selected definitions drawn from Biotechnology at Work.

Monoclonal antibody Highly specific, purified antibody that is derived from only one clone of cells and recognizes only one antigen.

Morphology The form and structure of plants and animals.

Nucleotide sequence The sequence of the bases within the nucleic acid that determines what proteins will be made. Nucleotides are the building blocks of nucleic acids (DNA and RNA). Each nucleotide is composed of sugar, phosphate, and one of four nitrogen bases.

Peptide Two or more amino acids bonded together.

Phenotype The visible characteristics (traits) of a plant or animal.

Plasmid A free-floating circular piece of DNA, not part of the main chromosome, in bacteria. It is capable of replicating independently in a host cell.

Protoplast The cellular material that remains after the cell wall has been removed.

PST Porcine somatotropin. The version of somatotropin produced by swine.

Regeneration The process of regenerating (growing) whole plants from cells taken from plants or plant parts and manipulated in tissue culture.

Restriction enzymes Enzymes that break DNA in highly specific locations, creating gaps into which new genes can be inserted.

Saprophyte An organism that obtains nutrients from dead or decaying organic matter.

Somaclonal variation The genetic changes that occur within non-reproductive cells, i.e., not via genetic engineering.

Somatotropin A protein hormone secreted by the pituitary gland in mammals. It directs lactation and growth. Each species produces its own specific version that is active in lower but not higher order species. Since it is a protein, somatotropin must be injected to be active.

Starter culture A culture of a known organism which is used to perform industrial fermentations.

Substrate Material acted on by an enzyme.

Tissue culture The suspension of plant cells in a nutrient bath to alter them genetically or coax them into producing whole plants through regeneration.

Transgenic organism An organism formed by the insertion of foreign genetic material into the reproductive cells.

Vector The agent (e.g., plasmid, virus) used to carry new DNA into a cell.

Authors

Randolph Barker
Professor
Department of Agricultural Economics
Director of the Southeast Asia Program
Cornell University

Bill R. Baumgardt
Director
Agricultural Experiment Station
Purdue University

James N. BeMiller
Professor
Department of Food Science
Purdue University

Lawrence M. Busch
Professor
Department of Sociology
Michigan State University

L. J. (Bees) Butler
Cooperative Extension Economist
Department of Agricultural Economics
University of California, Davis

Neville P. Clarke
Associate Deputy Chancellor for Agriculture
Texas A&M University

Michael P. Doyle
Professor
Food Research Institute
University of Wisconsin

James I. Grieshop
Specialist in Community Education
Department of Applied Behavioral Sciences
University of California, Davis

Susan K. Harlander
Associate Professor
Department of Food Science and Nutrition
University of Minnesota

Charles E. Hess
Assistant Secretary for Science & Education
United States Department of Agriculture

Glenn L. Johnson
Professor Emeritus
Department of Agricultural Economics
Michigan State University

Laura R. Lacy
Research Associate
Department of Plant Pathology
Pennsylvania State University.

William B. Lacy
Assistant Dean for Research
Professor of Rural Sociology
Pennsylvania State University

Victor L. Lechtenberg
Executive Associate Dean of Agriculture
Purdue University

David R. MacKenzie
Director
National Biological Impact Assessment Program
United States Department of Agriculture/ Cooperative State Research Service

James J. Marois
Associate Professor
Department of Plant Pathology
University of California, Davis

Elmer H. Marth
Emeritus Professor
Department of Food Science
University of Wisconsin

Marshall A. Martin
Professor
Department of Agricultural Economics
Purdue University

Joseph J. Molnar
Alumni Professor
Department of Agricultural Economics and Rural Sociology
Auburn University

Donald L. Plucknett
Science Advisor
Consultative Group on International Agricultural Research Secretariat
The World Bank

A. Allan Schmid
Professor
Department of Agricultural Economics
Michigan State University

Larry R. Steenson
Assistant Professor
Department of Food Science
Purdue University

W. Burt Sundquist
Professor
Department of Agricultural and Applied Economics
University of Minnesota

Paul B. Thompson
Professor
Department of Agricultural Economics
Professor of Philosophy
Director of the Center of Biotechnology Policy and Ethics
Texas A&M University

Anne K. Vidaver
Head and Professor
Department of Plant Pathology
University of Nebraska-Lincoln

Reviewers

Charles B. Browning
Oklahoma State University

Frederick H. Buttel
Cornell University

Charles L. (Larry) Chrisman
Purdue University

Donald N. Duvick
Pioneer Hi-Bred International

Beverly Fleisher
National Science Foundation

Lowell S. Hardin
Purdue University

Chuck Hassebrook
Center for Rural Affairs

Thomas J. Hoban
North Carolina State University

Thomas K. Hodges
Purdue University

Daniel D. Jones
USDA

A. David Kline
Iowa State University

Fred Kuchler
USDA/ERS

Stephen B. Lovejoy
Purdue University

Richard L. Lower
University of Wisconsin

David R. MacKenzie
NBIAP, USDA/CSRS

Terry L. Medley
USDA/APHIS

Paul O'Connell
USDA

Michael Olexa
USDA/ES

John Reilly
USDA/ERS

Joseph C. Roetheli
USDA/CSRS

Vernon W. Ruttan
University of Minnesota

A. Ann Sorensen
American Farm Bureau Federation

Martha Bair Steinbock
USDA

Paul E. Stern
University of Florida

Eileen O. van Ravenswaay
Michigan State University

Roger Wyse
Rutgers University

The NASULGC Division of Agriculture

Committee on Biotechnology

Charles B. Browning, Chairman
Dean and Director
Division of Agriculture
Oklahoma State University
Stillwater, OK

Bill R. Baumgardt
Director of Agricultural Research
Purdue University
West Lafayette, IN

Neville P. Clarke
Associate Deputy Chancellor
Texas A&M University
College Station, TX

James M. Davidson
Dean for Research
Institute of Food and Agricultural Sciences
University of Florida
Gainesville, FL

Richard L. Lower
Associate Dean
College of Agriculture and Life Sciences
University of Wisconsin
Madison, WI

David R. MacKenzie
Director,
National Biological Impact Assessment Program
USDA/CSRS
Washington, DC

Marshall A. Martin
Professor
Dept. of Agricultural Economics
Purdue University
W. Lafayette, IN

Paul E. Stern
Biotechnology Policy Advisor
Institute of Food and Agricultural Sciences
University of Florida
Gainesville, FL

T. Ross Wilkinson
Dean and Director of Resident Instruction
College of Agricultural Sciences
Clemson University
Clemson, SC

Socioeconomic Subcommittee

Members from Committee on Biotechnology

Bill R. Baumgardt
(Co-Chairman of Subcommittee)
Purdue University

Stephen J. Kleinschuster*
Rutgers University

Richard L. Lower
University of Wisconsin-Madison

Marshall A. Martin
(Co-Chairman of Subcommittee)
Purdue University

Wally J. Moline*
Michigan State University

Ad Hoc Members

Frederick H. Buttel
Cornell University

Susan E. Offutt
Office of Management and Budget

Michael J. Phillips
Office of Technology Assessment

Vernon W. Ruttan
University of Minnesota

Peter Sandman
Rutgers University

* *Members of Committee on Biotechnology when the Socioeconomic Subcommittee was instituted.*

The Committee on Biotechnology was appointed by the Chairman of the Division of Agriculture of the National Association of State Universities and Land-Grant Colleges in April 1982. The concept for this basebook and accompanying leaflet series was developed by a subcommittee assigned to deal with legal, ethical, and socioeconomic issues related to application of the new biological research techniques to agriculture.